Solutions Manual
for
Modern Genetic Analysis

William D. Fixsen

W. H. Freeman and Company
New York

ISBN 0-7167-3282-3

Printed in the United States of America

First printing 1998

CONTENTS

PREFACE

It is my hope that this manual will provide useful explanations, numerous approaches to solving problems, and of course, correct answers. However, having taught genetics for many years to thousands of students, it is my belief that focusing too much on the latter misses the point of the exercise. In the end, learning to recognize what steps you actually take, what knowledge and logic you apply, is more important than the answer. Certainly getting the correct answer is satisfying and should be considered the reward for all your hard work, but understanding how you solved the problem will ultimately be more useful. So when attempting to solve a problem, do not give up too easily and do not fool yourself by just looking at the answer and thinking "right, that *would* have been my answer too."

During my own education, I have had a number of teachers and mentors who have passed on their knowledge and love for genetics to whom I am indebted. Dr. Elizabeth Jones, Dr. David Botstein, Dr. Maurice Fox, and Dr. H. Robert Horvitz have all played key roles in my genetic training; I hope that I am able to convey some of their enthusiasm in my own work. I would also like to thank Dr. William Gelbart for recommending me for this project. Although I have worked with several people at W. H. Freeman, I would especially like to thank Jodi Isman for bringing this project to fruition.

William D. Fixsen
Department of Molecular and Cellular Biology
Harvard University

To my grandest genetic experiments,
Allie and Jessie

1 Genetics and the Organism

1. *Genetics* is the study of genes and genomes: their biochemical basis, how they function, how they are controlled, how they are organized, how they replicate, how they change, how they can be manipulated, and how they are transmitted from cell to cell and generation to generation.

   The ancient Egyptian racehorse breeders can be only loosely classified as geneticists because their interests were highly focused on producing fast horses, rather than on attempting to understand the mechanisms of heredity. Their understanding of the processes involved in producing fast horses was quite incorrect and their methods were not analytic in the modern sense of the word. Nevertheless, they did produce very fast horses through a combination of observation, trial and error, and artificial selection.

2. Genetics has affected modern society in law (copyrights, paternity suits, criminal cases), medicine in areas too numerous to list, agriculture, and virtually any other general area that exists. For example, agricultural pest control makes use of mutations to produce sterile males who then mate with females, producing no new generation. Genes from bacteria can be put into plants to make them resistant to various insects. Gene therapy offers potential cures for many human diseases. Pigs are being genetically engineered to someday produce organs for human transplant. The limit seems to be the imagination of those geneticists who have a focus on the practical uses of the ever-expanding base of genetic knowledge.

3. DNA determines all the specific attributes of a species (shape, size, form, behavioral characteristics, biochemical processes, etc.) and sets the limits for possible variation that is environmentally induced.

4. Properties of DNA that are vital to its being the hereditary molecule are: its ability to replicate, its informational content, and its relative stability while still retaining the ability to change or mutate. Alien life forms might utilize RNA, just as some viruses do, as a hereditary molecule. However, of the types of molecules that can exist on earth, only the nucleic acids possess the necessary characteristics.

5. The norm of reaction is the phenotypic variation that exists for a species of a fixed genotype within a varying environment. The variability itself can become vital to a species with a change in environment, for it allows for the possibility that some individuals may be able to survive under the new environmental conditions long enough to reproduce.

6. Phenotypic variation within a species can be due to genotype, environmental effects, and pure chance (random noise). Showing that variation of a particular trait has a genetic basis, especially for traits that show continuous variation, therefore requires very carefully controlled analyses. This is discussed in detail in Chapter 18 of the companion text.

7. The formula *genotype* + *environment* = *phenotype* is both accurate and inaccurate simultaneously. While phenotype is a product of the genotype and the environment, there is also an inherent variation due to random noise. Given complete information regarding both genotype and the environment, it is still impossible to specify the phenotype completely, although a close approximation can be achieved.

8. Following are five ethical dilemmas in genetics: (1) Attempting to deal with issues of world hunger, the Green Revolution was the result of using genetic knowledge and manipulation to generate higher yielding crops. Currently, plants are being genetically engineered for higher yield, changes in protein content, and resistance to insects, disease, and herbicides. However, it can also be argued that this has led to cultural disruption as farming techniques have been forced to adapt to the requirements of the seeds used, such as the extensive and expensive use of fertilizers, herbicides, and insecticides. There are also concerns that the monoculture of crops leads to loss of genetic diversity or that the genes from other organisms being inserted into the genomes of our food crops will result in unexpected food allergies, or changes in food quality. (2) What does an individual do with the knowledge (discovered through genetic testing) that she or he may die in 10 or 20 years from a genetic disorder that has yet to produce any symptoms and has no cure? Does the individual even want such knowledge? (3) What does society do with the knowledge that an individual is carrying a genetic disorder and is reproducing, and that the society will ultimately have to take care of the offspring of that person? (4) How can society protect the rights of an individual to privacy as more and more is known about that individual genetically? Does society have the right to prohibit reproduction? Do insurance companies have the right to the information about an individual whom they may insure? Does the individual have the right not to know of a genetic problem? (5) Genetic knowledge has been utilized in our legal system for decades, for example, in establishing paternity, or more recently, in using DNA "fingerprints" to place

suspects at the crime scene. But, there have been laboratory errors resulting in indictment of the innocent and exculpation of the guilty. In addition, juries are usually confused by the details of genetic evidence so that they very frequently either ignore it completely or accept it without any questioning.

9. The goal of genetic dissection is to identify, study, and understand the function and interaction of all the genes (and their products) that affect a specific trait.

10. All life uses uniform genetic systems. The study of any organism or life process can be compared to all others because, ultimately, DNA is the common thread that connects all. For example, evolution makes no sense without understanding how DNA changes or how genes lead to phenotype. Embryonic development is, in essence, genes in action. Each area of biology ultimately explains the details of its discipline, at least in part, with genetic knowledge.

2 The Structure of Genes and Genomes

1. There are a number of ways of connecting these terms. One might be:

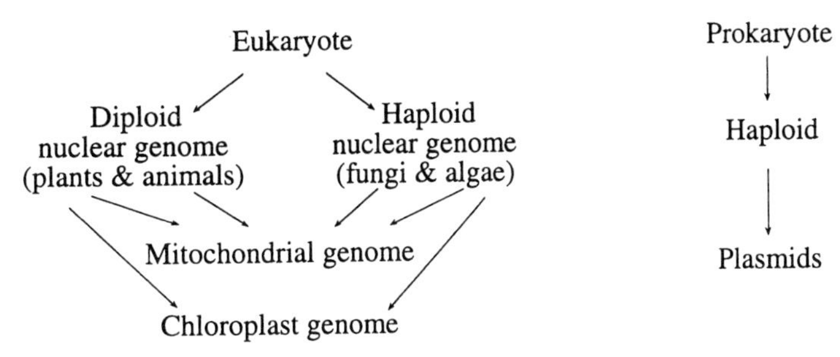

2. In each cell there are two copies of the nuclear genome (plants are diploid) and many copies of the mitochondrial and chloroplast genomes.

3. Diploid organisms have two copies of each of their different chromosomes. The term homologous is used to refer to the two members of each pair. Homologous chromosomes are alike in that they carry the same genes in the same relative positions, however, these genes may differ in informational content (representing different alleles).

4. **a.** $3 \times 10^9/1000 = 3 \times 10^6$ base pairs; **b.** 1 m/1000 = 1 mm

5. **a.** The following have mitochondria: a fish, moss, a palm tree, and bakers' yeast.

   b. The following have chloroplasts: a diatom and mistletoe.

6. The DNA double helix is held together by two types of bonds—covalent and hydrogen. Covalent bonds occur within each linear strand and strongly bond the bases, sugars, and phosphate groups (both within each component and between components). Hydrogen bonds occur between the two strands and involve a base from one strand with a base from the second in complementary pairing. These hydrogen bonds are individually weak but collectively quite strong.

7. If the DNA is double stranded, A = T and G = C and A + T + C + G = 100%. If T = 15%, then C = [100 – 15(2)]/2 = 35%.

8. If the DNA is double stranded, G = C = 24% and A = T = 26%.

9. There are many ways to indicate the polarity of DNA in a simple way. The point of this exercise is to realize that the polarity is based on how the sugar (deoxyribose) is oriented within the backbone of each strand. The 5′ carbon is attached to a phosphate while the 3′ carbon has an hydroxyl group to which new nucleotides may be added. In double-stranded DNA, the two strands are antiparallel, meaning that the sugars of each are oriented in opposite directions.

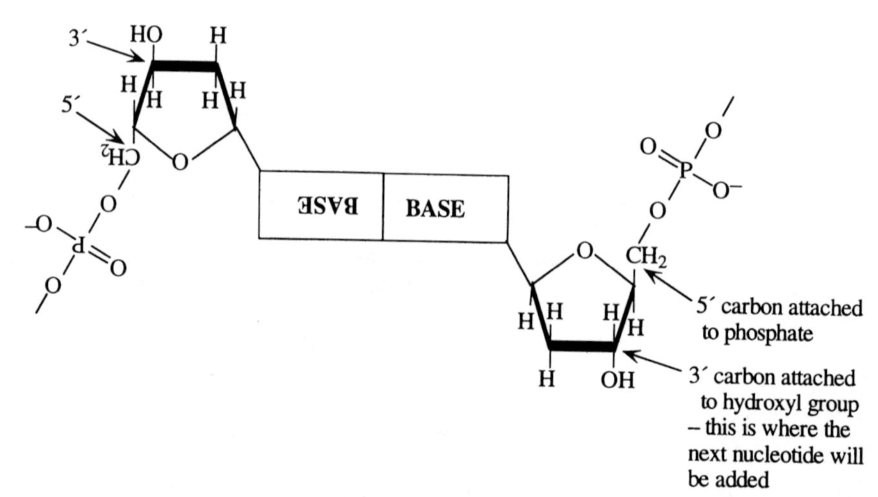

10. Human somatic cells are diploid (2*n*) and the haploid number is 23 (*n* = 23).

11. 46 DNA molecules (chromosomes)

12. When chromosomes become thick enough to be visible by light microscopy during cell division, the DNA has undergone remarkable packaging. Starting from "naked" DNA: it is first wrapped twice around cores of histone proteins called nucleosomes; the DNA/nucleosome complex is then coiled into a solenoid stabilized by another type of histone; this solenoid is arranged in loops and then supercoiled onto a core of nonhistone proteins called the scaffold.

13. 3′ – T A A C C G A G A – 5′

14. **a.** Yes. Since A = T and G = C, the equation A + C = G + T can be rewritten as T + C = C + T by substituting the equal terms.

b. Yes, the percentage of purines will equal the percentage of pyrimidines in double-stranded DNA.

15. Sulfur

16. The simplest definition is that a *gene* is a chromosomal region capable of making a functional transcript. However, this does not take into account the regulatory regions near the gene necessary for the proper expression of the gene or the regions that help control transcription that can be quite distant. Also, many eukaryotes have large regions of noncoding sequences (introns) interspersed within the regions that encode a product (exons).

17. a. 1,830 kb/1,703 genes = 1,074 base pairs

b. 1,074 – 1,000 = 74 base pairs

c. These DNA sequences will predominantly be regulatory regions.

d. 0%: Bacteria do not have introns.

e. The same as part b, 74 base pairs.

18. Many eukaryotes have intervening sequences called *introns* which are transcribed but then spliced out (removed) prior to translation. In this case, the 25-kb primary transcript has all but 2.1 kb removed to generate the proper mRNA necessary for translation.

19. PFGE separates DNA molecules by size. When DNA is carefully isolated from *Neurospora* (which has 7 different chromosomes) 7 bands should be produced using this technique. Similarly, the pea has 7 different chromosomes and will produce 7 bands (homologous chromosomes will co-migrate as a single band). The housefly has 6 different chromosomes and should produce 6 bands.

20. mRNA size = gene size – (number of introns × average size of introns)

21. (1.7%) $3 \times 10^9 = 51 \times 10^6$ nucleotide differences

22. a. The purine bases are the larger, two-ring (planar, aromatic, heterocyclic) structures while the pyrimidines are the smaller, one-ring structures. Thus each purine/pyrimidine base pair is of equal width.

b. Adenine and guanine are both purines, however the chemical structure of guanine (with one keto and one amino group) allows for three hydrogen bonds when paired with cytosine while adenine (with one amino group) forms two hydrogen bonds when paired with thymine.

c. Cytosine and thymine are both pyrimidines, however thymine has one methyl and two keto groups on its ring while cytosine has one amino and one keto group on its ring.

23. The polarity of the sugar-phosphate backbone in DNA is defined by the orientation of the deoxyribose sugar. By convention, the carbons are numbered as primes to differentiate the atoms of the base from those of the sugar. It is useful to use 5′ and 3′ to define the orientation of DNA (or RNA) strands. In this way, the two strands of the double helix are said to be antiparallel (run in opposite orientations). Also, as you learn more about the enzymology of replication, transcription, and translation, the importance and concept of strand orientation can easily be conveyed.

24.

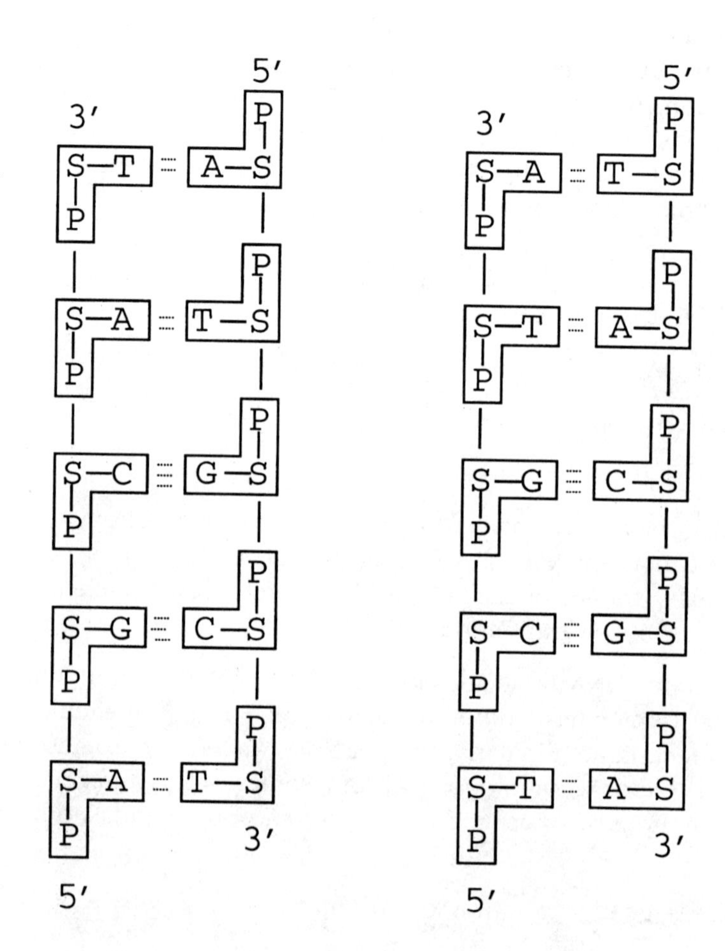

25. 2 kb = 2000 bp = 4000 bases

26. There are roughly 10 base pairs per turn.

27. 9 polytenes (each homologous pair would be joined as one).

3 Gene Function

1. There are three codons for isoleucine: 5´ AUU 3´, 5´ AUC 3´, and 5´ AUA 3´. Possible anticodons are 3´ UAA 5´ (complementary), 3´ UAG 5´ (complementary), and 3´ UAI 5´ (wobble). 5´ UAU 3´, although complementary, would also base pair with 5´ AUG 3´ (methionine) due to wobble and therefore would not be an acceptable alternative.

2. **a.** By studying the genetic code table provided in the textbook, you will discover that there are eight cases in which knowing the first two nucleotides does not tell you the specific amino acid.

b. If you knew the amino acid, you would not know the first two nucleotides in the cases of Arg, Ser, and Leu.

3.

3´ C G T	A C C	A C T	G C A 5´	DNA double helix (transcribed strand)
5´ G C A	T G G	T G A	C G T 3´	DNA double helix
5´ G C A	U G G	U G A	C G U 3´	mRNA transcribed
3´ C G U	A C C	A C U	G C A 5´	Appropriate tRNA anticodon
NH_3 - Ala - Trp - (stop) - COOH				Amino acids incorporated

4. **a., b.** 5´ UUG GGA AGC 3´

c., d. Assuming the reading frame starts at the first base:
NH_3 - Leu - Gly - Ser - COOH

For the bottom strand, the mRNA is 5´ GCU UCC CAA 3´ and assuming the reading frame starts at the first base, the corresponding amino acid chain is NH_3 - Ala - Ser - Gln - COOH.

5. **a.** The data cannot indicate whether one or both strands are used for transcription. You do not know how much of the DNA is transcribed nor which regions of DNA are transcribed.

b. If the RNA is double stranded, the percentage of purines (A + G) would equal the percentage of pyrimidines (U + C) and the (A +G)/(U + C) ratio would be 1.0. This is clearly not the case for *E. coli*, which has a ratio of 0.80. The ratio for *B. subtilis* is 1.02. This is consistent with the RNA being double stranded but does not rule out single stranded if there are an equal number of purines and pyrimidines in the strand.

6. **a.** The main use is in detecting carrier parents and in diagnosing the disorder in the fetus.

b. Because the values for normal individuals and carriers overlap for galactosemia, there is ambiguity if a person has 25 to 30 units. That person could be either a carrier or normal.

c. These wild-type genes are phenotypically dominant but are incompletely dominant at the molecular level. A minimal level of enzyme activity apparently is enough to ensure normal function and phenotype.

7. Growth will be supported by a particular compound if it is later in the pathway than the enzymatic step blocked in the mutant. Restated, the more mutants a compound supports, the later in the pathway it must be. In this example, compound G supports growth of all mutants and can be considered the end product of the pathway. Alternatively, compound E does not support the growth of any mutant and can be considered the starting substrate for the pathway. The data indicate the following:

a., b.

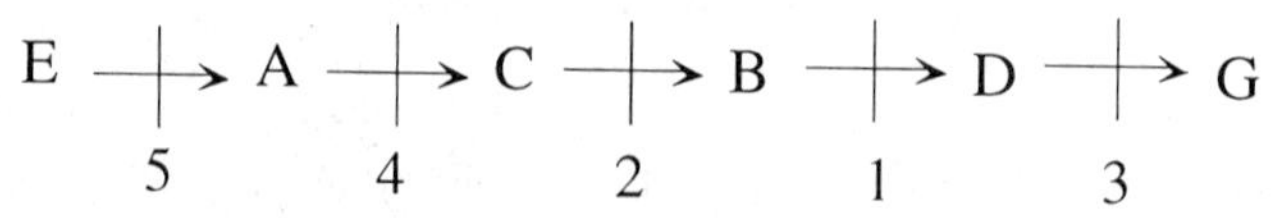

E —|→ A —|→ C —|→ B —|→ D —|→ G
5 4 2 1 3

Vertical lines indicate step where each mutant is blocked.

8. **a.** If enzyme A was defective or missing (*m2/m2*), red pigment would still be made and the petals would be red.

b. Purple, because it has a wild-type allele for each gene and you are told that the mutations are recessive.

c. The mutant alleles do not produce functional enzyme. However, enough functional enzyme must be produced by the single wild-type allele of each gene to synthesize normal levels of pigment.

9. **a.** If enzyme B is missing, a white intermediate will accumulate and the petals will be white.

b. If enzyme D is missing, a blue intermediate will accumulate and the petals will be blue.

10. There are a number of explanations that might explain the positional clustering of null mutants in this *Neurospora* gene. However, based on the material covered in this chapter, it is possible that the gene codes for an enzyme that uses its central amino acids (encoded by the central region of the gene) to form its active site. Mutations that alter the active site amino acids are most likely to destroy enzymatic function and would be classified as null (loss-of-function) mutations.

11. a. Null alleles are the result of mutations that destroy the function of the protein. Missense, nonsense, deletion, and frameshift mutations within exons may all result in loss-of-function alleles. Although transcribed, the polypeptide may or may not be translated (or recognized by the experimenter) depending on the type of mutation.

b. There are sequences near the boundaries of and within introns that are necessary for correct splicing. If these are altered by mutation, correct splicing will be disrupted. Although transcribed, it is likely that translation will not occur.

c. The promoter is required for the proper initiation of transcription. Null alleles of the promoter will block transcription.

d. Yes; transcript yes; polypeptide no (see part b).

12. Protein function can be destroyed by a mutation that causes the substitution of a single amino acid even though the protein has the same immunological properties. For example, enzymes require very specific amino acids in exact positions within their active site. A substitution of one of these key amino acids might have no effect on overall size and shape of the protein while completely destroying the enzymatic activity.

13. a. Each wild-type allele produces 10 units of enzyme E per cell. A null mutation in the gene coding for this enzyme produces 0 units. More than 10 units of enzyme are required by the cells for wild-type levels of function and phenotype.

E^+/E^+	20 units	Wild type
E^+/E	10 units	Mutant
E/E	0 units	Mutant

b. Each wild-type allele produces 15 units of enzyme F per cell. A null mutation in the gene coding for this enzyme produces 0 units. 15 units of enzyme are sufficient for wild-type phenotype.

f^+/f^+	30 units	Wild type
f^+/f	15 units	Wild type
f/f	0 units	Mutant

c. Each wild-type allele (g^+) produces 25 units of enzyme G per allele. Mutant 1 (g^1) is leaky, producing 5 units of enzyme G per allele and mutant 2 (g^2) is a null mutant producing 0 units per allele.

14. a. The allele *sn* will show dominance over *sf* because there will be only 40 units of square factor in the heterozygote.

b. No. Here the functional allele is recessive.

c. The allele *sf* may become dominant over time by several mechanisms: it could mutate in a way that it produces 50 units; other genes might mutate to lower the cells need to have 50 units; or other genes might mutate to increase the production or activity of *sf*.

15. a. The mothers had an excess of phenylalanine in their blood, and that excess was passed through the placenta into the fetal circulatory system, where it caused brain damage prior to birth.

b. The diet had no effect because the neurological damage has already happened *in utero* prior to birth.

c. A fetus with two mutant copies of the allele that causes PKU makes no functional enzyme. However, the mother of such a child is heterozygous and makes enough enzyme to block any brain damage; the excess phenylalanine in the fetal circulatory system enters the maternal circulatory system and is processed by the maternal gene product. After birth, which is when PKU damage occurs in a PKU child, dietary restrictions block a buildup of phenylalanine in the circulatory system until brain development is completed.

The fetus of a PKU mother is exposed to the very high level of phenylalanine in its circulatory system during the time of major brain development. Therefore, brain damage occurs before birth, and no dietary restrictions after birth can repair that damage.

d. The obvious solution to the brain damage seen in the babies of PKU mothers is to return the mother to a restricted diet during pregnancy in order to block high levels of exposure to her child.

e. PKU is characterized as a rare recessive disorder. A child with PKU has two parents that carry a mutant allele for the metabolism of phenylalanine. When two individuals who are heterozygous for PKU have a child, the risk that the child will have PKU is 25%. A PKU child is unable to make a functional enzyme that converts phenylalanine to tyrosine. As a result, an excess level of phenylalanine is found in the blood, and the excess is detected as an increase in phenylpyruvic acid in both the blood and the urine. The excess phenylpyruvic acid blocks normal development of the brain, resulting in retardation.

16. a., b. The goal of this type of problem is to align the two sequences. You are told that there is a single nucleotide addition and single nucleotide deletion, so look for single base differences that effect this alignment. These should be located where the protein sequence changes (i.e., between Lys–Ser and Asn–Ala). Remember also that the genetic code is redundant. (N = any base)

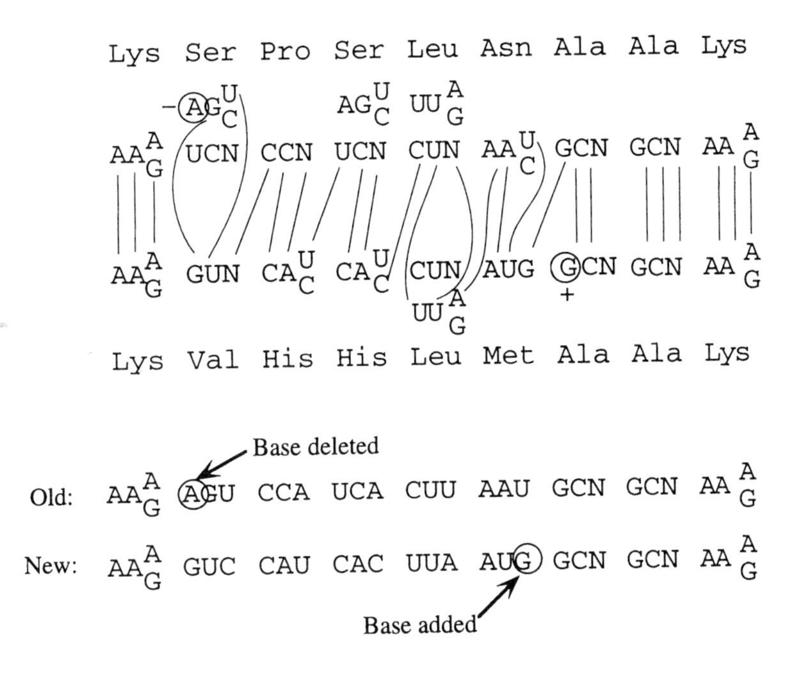

17. **a.** Recessive. The normal allele provides enough enzyme to be sufficient for normal function (the definition of *haplo-sufficient*).

b. There are many ways to mutate a gene to destroy enzyme function. One possible mutation might be a frameshift mutation within an exon of the gene. Assuming that a single base pair was deleted, the mutation would completely alter the translational product 3′ to the mutation.

c. Hormone replacement could be given to the patient.

d. If the hormone is required before birth, it can be supplied by the mother.

18. DNA has often been called the "blueprint of life" but how does it actually compare to a blueprint used for house construction? Both are abstract representations of instructions for building three-dimensional forms, and both require interpretation for their information to be useful. But real blueprints are two-dimensional renderings of the various views of the final structure drawn to scale. There is one-to-one correlation between the lines on the drawing and the real form. The information in the DNA is encoded in a linear array — a one-dimensional set of instructions that only becomes three-dimensional as the encoded linear array of amino acids fold into their many forms. Included in the informational content of the DNA are also all the directions required for its "house" to maintain and repair itself, respond to change, and replicate. Let's see a real blueprint do that!

19. Other than the human ability to synthesize and wear synthetic materials, living systems are "proteins or something that has been made by a protein."

20. Mutant 1: null mutation; no product is made.
Mutant 2: leaky mutation; not enough product is made for normal function.
Mutant 3: nonsense mutation/frameshift; a truncated protein product is made.
Mutant 4: substitution mutation; protein is the same size but a single amino acid change destroys its function.

21. **a.** *his*⁻ (since it cannot make histidine)

b. *y* (lowercase because it is recessive)

c. *Pr* (uppercase because it is dominant)

d. *H* (uppercase because it is dominant). Although most null mutations are recessive, some are dominant due to haplo-insufficiency.

e. y^+/y; Pr/Pr^+; H/H^+

22. Transcription (mRNA/DNA); translation (tRNA/mRNA); translation initiation (rRNA/mRNA); splicing (snRNP/mRNA)

23. There are many: polymerases, nucleases, transcription factors, ligases, restriction enzymes, ribosomal proteins, histones, topoisomerases, etc.

24. **a., b., c., d., e., f.**

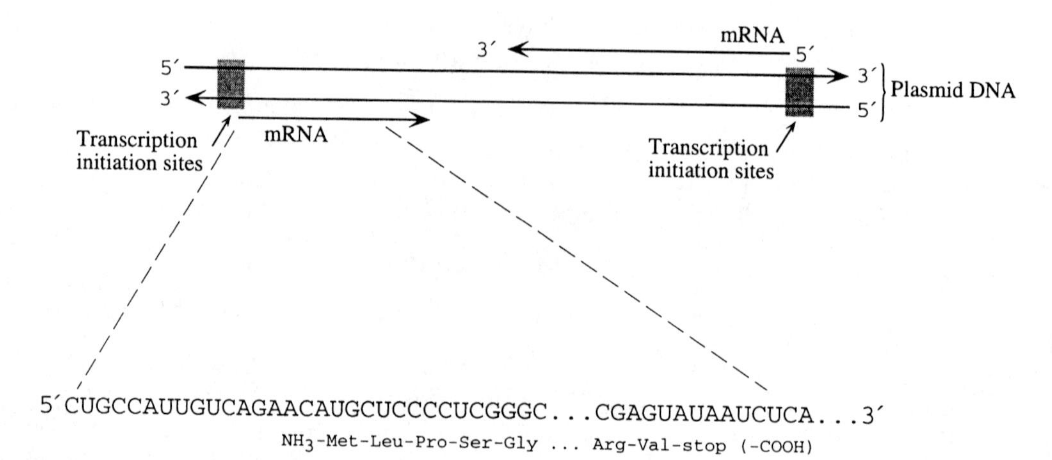

4 The Inheritance of Genes

1. The following drawing uses a solid line for the starting, nonradioactive DNA and a dotted line for the newly synthesized, radioactive DNA. The arrow heads are used to indicate the 3´ direction of the DNA strand, and the oval represents the chromosome's centromere. For simplicity, a single chromosome is shown even though these cells are diploid. In (d), half the daughters will have DNA radioactive in both strands and half will have DNA that is half radioactive and half nonradioactive.

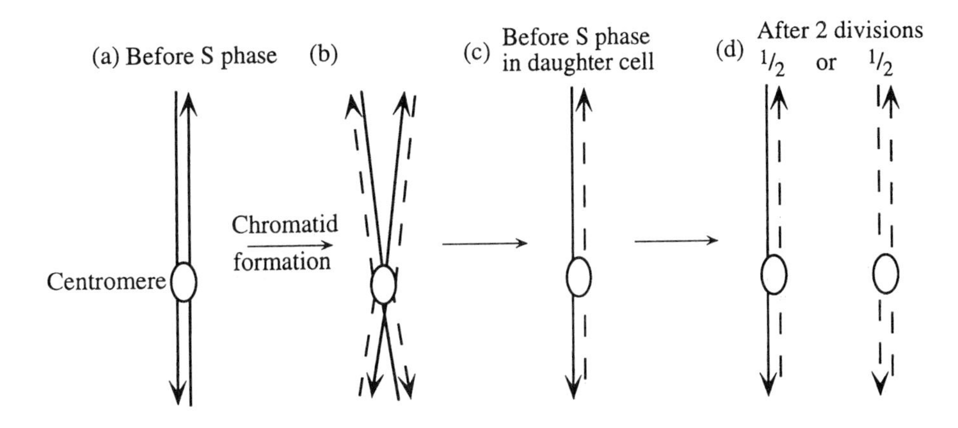

2. Graph of DNA content during mitosis and then meiosis

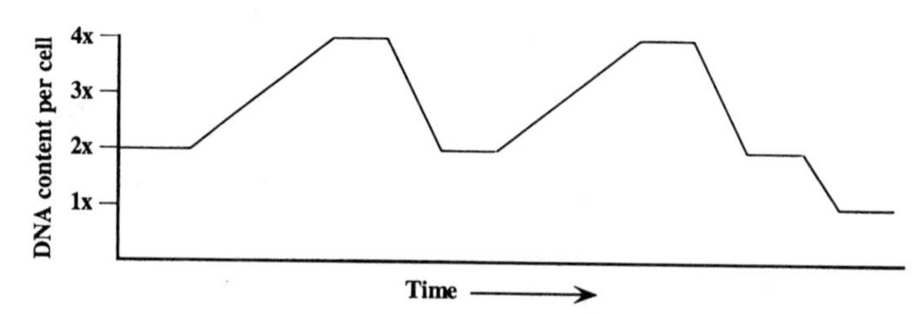

3. Because the DNA polymerase is capable of adding new nucleotides only at the 3´ end of a DNA strand, and because the two strands are antiparallel, at least two molecules of DNA polymerase must be involved in the replication of any specific region of DNA. When a region becomes single stranded, the two strands have an opposite orientation. Imagine a single-stranded region that runs from right to left. The 5´ end is at the right, with the 3´ end pointing to the left; synthesis can initiate and continue uninterrupted toward the right end of this strand. Remember: New nucleotides are added in a 5´→3´ direction, so the template must be copied from its 3´ end. The other strand has a 5´ end at the left with the 3´ end pointing right. Thus, the two strands are oriented in opposite directions (antiparallel), and synthesis (which is 5´→3´) must proceed in opposite directions. For the leading strand (say, the top strand) replication is to the right, following the replication fork. It is continuous and may be thought of as moving "downstream." Replication on the bottom strand cannot move in the direction of the fork (to the right), since, for this strand, that would mean adding nucleotides to its 5´ end. Therefore, this strand must replicate discontinuously: As the fork creates a new single-stranded stretch of DNA, this is replicated *to the left* (away from the direction of fork movement). For this lagging strand, the replication fork is always opening new single-stranded DNA for replication *upstream* of the previously replicated stretch, and a new fragment of DNA is replicated back to the previously created fragment. Thus, one (Okazaki) fragment follows the other in the direction of the replication fork, but each fragment is created in the opposite direction.

4. **a.** The bottom strand.

b.

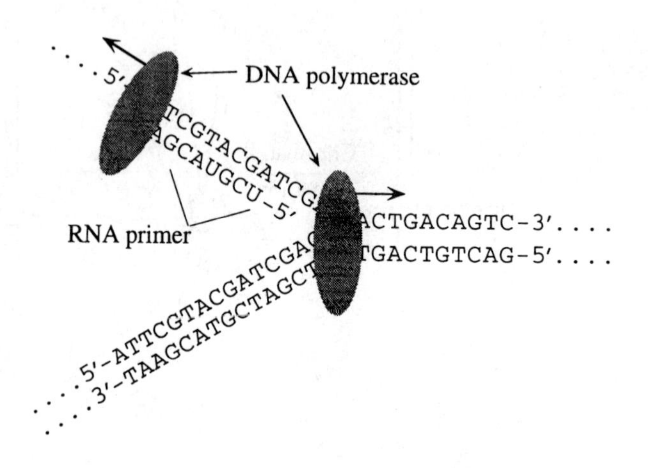

c.

```
....5'-ATTCGTACGATCGACTGACTGACAGTC-3'....
....3'-TAAGCATGCTAGCTGACTGACTGTCAG-5'....

....5'-ATTCGTACGATCGACTGACTGACAGTC-3'....
....3'-TAAGCATGCTAGCTGACTGACTGTCAG-5'....
```

d. Yes, but the other replication fork would be moving in the opposite direction; the top strand, as drawn, would now be the leading strand and the bottom strand would now be the lagging strand.

5. Horizontal lines (H) is dominant to vertical lines (h)

$h/h \times H/H$
↓
All H/h

Self $H/h \times H/h$
↓
$3/4$ ($1/4$ H/H + $1/2$ H/h)
$1/4$ h/h

6. Black (B) is dominant to white (b)

Parents $B/B \times B/b$
↓
Progeny $B/B \times B/b \times B/b \times B/B$

$B/B \times B/b$	$B/b \times B/b$	$B/b \times B/B$
$1/2$ B/b	$1/2$ B/b	$1/2$ B/B
$1/2$ B/B	$1/4$ B/B	$1/2$ B/b
	$1/4$ b/b	

7. Vertical line (h^+) is dominant to horizontal line (h)

$h/h \times h^+/h$
↓
$1/2$ h^+/h
$1/2$ h/h

$h/h \times h^+/h$ and $1/2$ h^+/h → $3/4$ $h^+/–$, $1/4$ h/h

8. Purple (B) is dominant to blue (b) and the trait is X-linked.

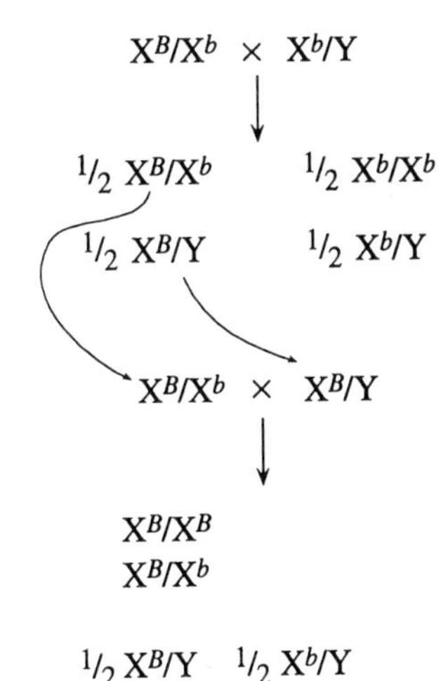

9. Star (s) is recessive to no star (s^+) and is X-linked.

$$X^s/X^s \times X^{s+}/Y$$
$$\downarrow$$
$$X^{s+}/X^s \times X^s/Y$$
$$\downarrow$$
$$^1/_2\ X^{s+}/X^s \qquad ^1/_2\ X^s/X^s$$
$$^1/_2\ X^{s+}/Y \qquad ^1/_2\ X^s/Y$$

10. Do a testcross (cross to *a/a*). If the fly was *A/A*, all the progeny will be phenotypically A; if the fly was *A/a*, half the progeny will be A and half will be a.

11. Black (*B*) is dominant to white (*b*)

Parents: *B/b* × *B/b*
Progeny: 3 black:1 white (1 *B/B*:2 *B/b*:1 *b/b*)

12. Charlie, his mate, or both, obviously were not pure breeding because his F_2 progeny were of two phenotypes. Let *A* = black and white, and *a* = red and white. If both parents were heterozygous, then red and white would have been expected in the F_1 generation. Red and white were not observed in the F_1 generation, so only one of the parents was heterozygous. The cross is:

P *A/a* × *A/A*
F_1 1 *A/a*:1 *A/A*

Two F_1 heterozygotes (*A/a*) when crossed would give 1 *A/A* (black and white):2 *A/a* (black and white):1 *a/a* (red and white). If the red and white F_2 progeny were from more than one mate of Charlie's, then the farmer acted correctly. However, if the F_2 progeny came only from one mate, the farmer may have acted too quickly.

13. The plants are approximately 3 blotched:1 unblotched. This suggests that blotched is dominant to unblotched and that the original plant which was selfed was a heterozygote.

a. Let *A* = blotched, *a* = unblotched.

P *A/a* (blotched) × *A/a* (blotched)
F_1 1 *A/A*:2 *A/a*:1 *a/a*
3 *A/–* (blotched):1 *a/a* (unblotched)

b. All unblotched plants should be pure-breeding in a testcross with an unblotched plant (*a/a*), and one-third of the blotched plants should be pure-breeding.

14. The results suggest that winged (*A/–*) is dominant to wingless (*a/a*) (cross 2 gives a 3:1 ratio). If that is correct, the crosses become

Pollination	Genotypes	Number of progeny plants	
		Winged	Wingless
Winged (selfed)	$A/A \times A/A$	91	1*
Winged (selfed)	$A/a \times A/a$	90	30
Wingless (selfed)	$a/a \times a/a$	4*	80
Winged × wingless	$A/A \times a/a$	161	0
Winged × wingless	$A/a \times a/a$	29	31
Winged × wingless	$A/A \times a/a$	46	0
Winged × winged	$A/A \times A/–$	44	0
Winged × winged	$A/A \times A/–$	24	0

The five unusual plants are most likely due either to human error in classification or to contamination. Alternatively, they could result from environmental effects on development. For example, too little water may have prevented the seed pods from becoming winged even though they are genetically winged.

15. a. The disorder appears to be dominant because all affected individuals have an affected parent. If the trait was recessive, then I-1, II-2, III-1, and III-8 would all have to be carriers (heterozygous for the rare allele).

b. Assuming dominance, the genotypes are
I: *d/d, D/d*
II: *D/d, d/d, D/d, d/d*
III: *d/d, D/d, d/d, D/d, d/d, d/d, D/d, d/d*
IV: *D/d, d/d, D/d, d/d, d/d, d/d, d/d, D/d, d/d*

c. The mating is *D/d* × *d/d*. The probability of an affected child (*D/d*) equals $1/2$, and the probability of an unaffected child (*d/d*) equals $1/2$. Therefore, the chance of having four unaffected children (since each is an independent event) is: $1/2 \times 1/2 \times 1/2 \times 1/2 = 1/16$.

16. a. Autosomal recessive: affected individuals inherited the trait from unaffected parents and a daughter inherited the trait from an unaffected father.

b. Both parents must be heterozygous to have a $1/4$ chance of having an affected child. Parent 2 is heterozygous since her father is homozygous for the recessive allele, and parent 1 has a $1/2$ chance of being heterozygous since his father is heterozygous because 1's paternal grandmother was affected. Overall, $1 \times 1/2 \times 1/4 = 1/8$.

17. a. leu^+/leu

b. The alleles will segregate during meiosis and the progeny will be:
1 leu^+:1 *leu*.

c. Certain amino acids are essential to protein structure and function, and a change of even one of these could totally destroy an enzyme's activity. There are many ways to change a DNA sequence that encodes an enzyme that will result in an altered amino acid sequence. For the following, a small DNA sequence has arbitrarily been chosen to show three classes of mutation that could cause loss of enzymatic activity.

The first, a missense mutation, is the result of a single base-pair change and alters a single amino acid. The second, a deletion of three base pairs, causes the loss of one amino acid. The third, a frameshift mutation caused by the addition of one base pair, alters the amino sequence from the site of the addition until the end of translation is reached. Other mutations, not shown, would also lead to null alleles: mutations in the promoter for this gene preventing transcription, mutations that alter or prevent proper splicing, insertions of DNA into the coding region of the gene, etc.

Wild-type allele

```
    ....5'-ATTCGTACGATCGAC-3'....
    ....3'-TAAGCATGCTAGCTG-5'....       DNA
          /                \
  ....5'-AUU CGU ACG AUC GAC-3'....       mRNA

....NH2-Ile Arg Thr Ile Asp-COOH....   Protein
```

Missense allele

```
    ....5'-ATTCATACGATCGAC-3'....
    ....3'-TAAGTATGCTAGCTG-5'....       DNA
          /                \
  ....5'-AUU CAU ACG AUC GAC-3'....       mRNA

....NH2-Ile His Thr Ile Asp-COOH....   Protein
```

Small deletion allele (3 pase pairs)

```
               CGT
               GCA
               \/
    ....5'-ATTACGATCGAC-3'....
    ....3'-TAATGCTAGCTG-5'....          DNA
          /            \
  ....5'-AUU ACG AUC GAC-3'....           mRNA

....NH2-Ile Thr Ile Asp-COOH....        Protein
```

Frameshift allele (+1)

```
    ....5'-ATTCGTAACGATCGAC-3'....
    ....3'-TAAGCATTGCTAGCTG-5'....      DNA
          /                  \
  ....5'-AUU CGU AAC GAU CGA C-3'....   mRNA

....NH2-Ile Arg Asn Asp Arg-COOH....   Protein
```

18. P s^+/s^+ × s/Y
↓

F_1 $^1/_2$ s^+/s normal female
$^1/_2$ s^+/Y normal male

s^+/s × s^+/Y
↓

F_2 $^1/_4$ s^+/s^+ normal female
$^1/_4$ s^+/s normal female
$^1/_4$ s^+/Y normal male
$^1/_4$ s/Y small male

In the cross: P s^+/s × s/Y
↓

Progeny $^1/_4$ s^+/s normal female
$^1/_4$ s/s small female
$^1/_4$ s^+/Y normal male
$^1/_4$ s/Y small male

19. You should draw pedigrees for this question.

a.

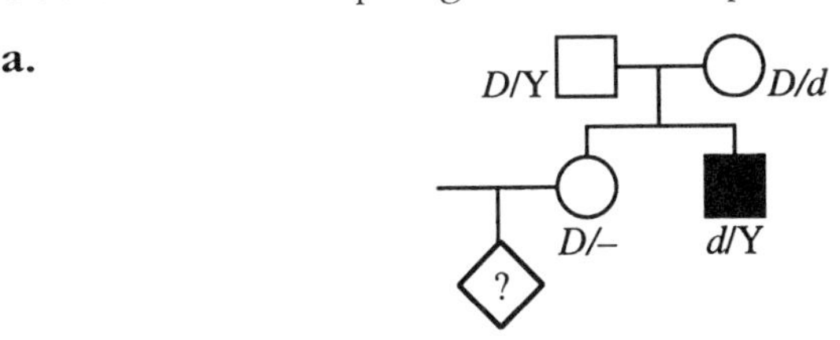

The "maternal grandmother" had to be a carrier, *D/d*. The probability that the woman inherited the *d* allele from her is $^1/_2$. The probability that she passes it to her child is $^1/_2$. The probability that the child is male is $^1/_2$. The total probability of the woman having an affected child is $^1/_2 \times {}^1/_2 \times {}^1/_2 = {}^1/_8$.

b. The same pedigree as part a applies. The "maternal grandmother" had to be a carrier, *D/d*. The probability that your mother received the allele is $^1/_2$. The probability that your mother passed it to you is $^1/_2$. The total probability is $^1/_2 \times {}^1/_2 = {}^1/_4$.

c.

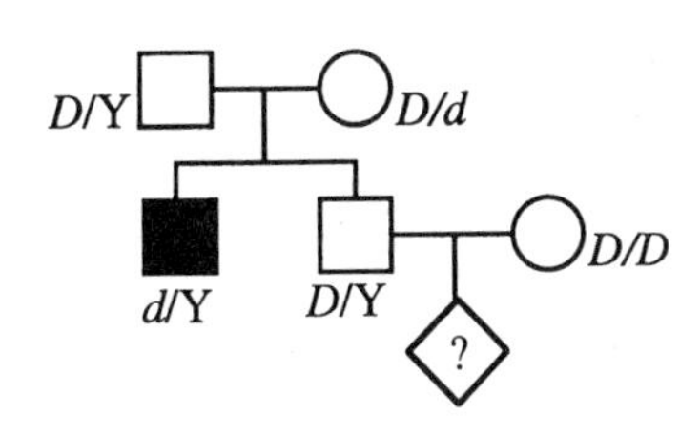

Because your father does not have the disease, you cannot inherit the allele from him. Therefore, the probability of inheriting an allele will be based on the chance that your mother is heterozygous. Since she is "unrelated" to the pedigree, assume that this is zero.

20. a. The pedigree indicates that the disease is caused by an X-linked recessive allele. You are told that the trait is rare, and in this pedigree, only males are affected. Since both affected males have unaffected parents, the allele must be recessive. If autosomal, both parents of both affected males would have to be heterozygous, but since it is rare, it is more likely that the grandmother was heterozygous for an X-linked recessive allele that was inherited by one son and the second daughter, who subsequently passed the allele to one of her sons.

b.

Generation I: X^+/Y, X^+/X^m
Generation II: X^+/X^+, X^m/Y, X^+/Y, $X^+/–$, X^+/X^m, X^+/Y
Generation III: X^+/X^+, X^+/Y, X^+/X^m, X^+/X^m, X^+/Y, X^+/X^+, X^m/Y, X^+/Y, $X^+/–$

c. The first couple has little chance of having an affected child because the male is normal, and the chance of the unrelated female's being heterozygous is rare. The second couple has a 50% chance of having affected sons and no chance of having affected daughters. The third couple has little chance of having an affected child (again, because the chance of the unrelated female's being heterozygous is rare), but all daughters will be heterozygous.

21. a. From cross 6, Bent (*B*) is dominant to normal (*b*). Both parents are "bent," yet some progeny are "normal."

b. From cross 1, it is X-linked. The trait is inherited in a sex-specific manner — all sons have the mother's phenotype.

c. In the following table, the Y chromosome is stated; the X is implied.

	Parents		Progeny	
Cross	Female	Male	Female	Male
1	*b/b*	*B*/Y	*B/b*	*b*/Y
2	*B/b*	*b*/Y	*B/b*, *b/b*	*B*/Y, *b*/Y
3	*B/B*	*b*/Y	*B/b*	*B*/Y
4	*b/b*	*b*/Y	*b/b*	*b*/Y
5	*B/B*	*B*/Y	*B/B*	*B*/Y
6	*B/b*	*B*/Y	*B/B*, *B/b*	*B*/Y, *b*/Y

22. a., b. The "stopper" or "continuous" phenotype of the maternal parent is inherited by all its offspring. This is typical of maternal inheritance (or organelle-based inheritance) and would be expected if this trait mapped to a mitochondrial gene.

Nuclear genes that differ between the two parental strains should segregate in the normal Mendelian manner and produce 1:1 ratios in the progeny. (Remember, *Neurospora* is haploid and the progeny of these crosses are actually the haploid products of meiosis.) This is observed for the orange/yellow trait suggesting that this trait maps to a nuclear gene.

23. a. Galactosemia pedigree

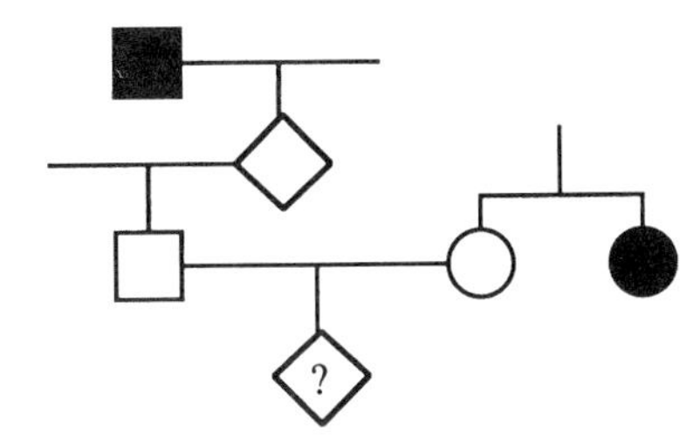

b. Both parents must be heterozygous for this child to have a $^1/_4$ chance of inheriting the disease. Since the mother's sister is affected with galactosemia, their parents must have both been heterozygous. Since the mother does not have the trait, there is a $^2/_3$ chance that she is a carrier (heterozygous). One of the father's parents must be a carrier since his grandfather had the recessive trait. Thus, the father had a $^1/_2$ chance of inheriting the allele from that parent. Since these are all independent events, the child's risk is:

$$^1/_4 \times {}^2/_3 \times {}^1/_2 = {}^1/_{12}$$

c. If the child has galactosemia, both parents must be carriers and thus those probabilities become 100%. Now all future children have a $^1/_4$ chance of inheriting the disease.

24. a. Favism pedigree

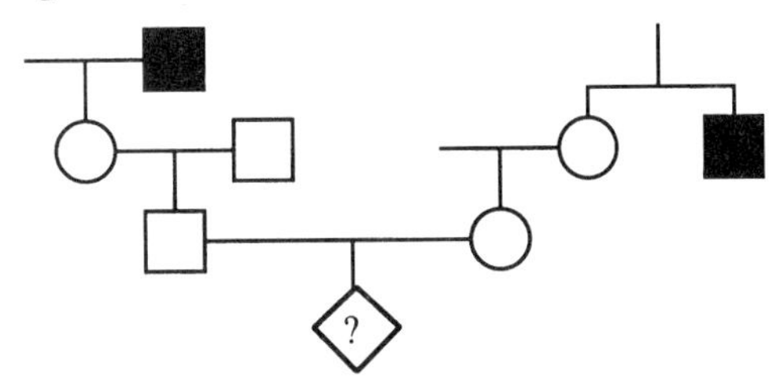

b. You are told that favism is due to an X-linked recessive allele. The father does not have favism so he must have the wild-type allele. Thus, for their child to inherit this trait, the mother must be a carrier (heterozygous) and the child must be male. The mother's uncle has favism, so we can assume that her maternal grandmother was heterozygous for the allele. In that case, there is a $^1/_2$ chance that her mother inherited the allele and another $^1/_2$ chance that she passed it to her. Since these are independent, there is a $^1/_4$ chance that the mother is a carrier. Finally, if she is a carrier, there is a $^1/_2$ chance she will pass the allele to her child and $^1/_2$ chance that the child is male:

$$^1/_4 \times {}^1/_2 \times {}^1/_2 = {}^1/_{16}$$

c. Once it is determined that the mother is a carrier, the chance of passing the allele to the next child is still $^1/_2$ as is the chance of that child being male, so $^1/_2 \times ^1/_2 = ^1/_4$.

25. a. Recessive. Affected individuals inherited the trait from unaffected parents.

b. Autosomal. A daughter inherited the trait from an unaffected father.

c. Both parents must be heterozygous for this child to have a $^1/_4$ chance of inheriting the disease. A's great-grandparents must both have been heterozygous while B's grandparents must both have been heterozygous. A's paternal grandmother is unaffected so she has a $^2/_3$ chance of being heterozygous, thus A's father has a $^2/_3 \times ^1/_2 = ^1/_3$ chance of being heterozygous and A has a $^2/_3 \times ^1/_2 \times ^1/_2 = ^1/_6$ chance of being heterozygous. Similarly, B's father has a $^2/_3$ chance of being heterozygous and B has a $^2/_3 \times ^1/_2 = ^1/_3$ chance of being heterozygous. Finally, the chance of an affected child is $^1/_4 \times ^1/_6 \times ^1/_3 = ^1/_{72}$.

26. *H* = hairy allele; *h* = smooth allele

plant 1: *H/h*
plant 2: *H/H*
plant 3: *H/h*
plant 4: *h/h*

Cross 2 indicates that hairy is dominant to smooth and both plant 1 and plant 3 are heterozygous. Plant 4 must be homozygous for the smooth allele since it exhibits the recessive trait, and plant 2 must be homozygous for the dominant allele since all progeny of plant 2 × plant 4 are hairy.

The progeny of cross 1 are $^1/_2$ *H/H* and $^1/_2$ *H/h;* of cross 2 are $^1/_4$ *H/H,* $^1/_2$ *H/h,* and $^1/_4$ *h/h;* of cross 3 are $^1/_2$ *H/h* and $^1/_2$ *h/h;* and of cross 4 all are *H/h.*

27.

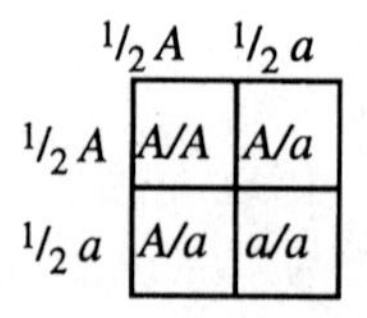

	$^1/_2$ *A*	$^1/_2$ *a*
$^1/_2$ *A*	*A/A*	*A/a*
$^1/_2$ *a*	*A/a*	*a/a*

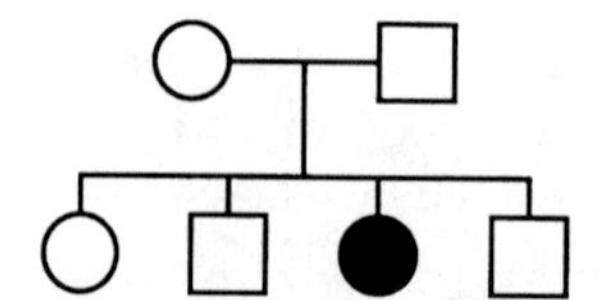

A pedigree is used to show genetic relationships between family members. Through deduction and interpretation of patterns, modes of inheritance may be indicated or ruled out. The genotypes of individuals represented in the pedigree can also be deduced in some cases. This information could prove useful both in understanding how genes affect phenotype and in providing genetic counseling. In a grid, a cross is represented as an abstract construct. It illustrates the probability of events that can happen but does not convey actual outcomes. Nonetheless, knowing that there is a $^1/_4$ chance of having a child that is *a/a* when both parents are *A/a* is useful information even if they already have four children that are not.

28. a. Pedigree 1: X-linked recessive

Affected individuals have inherited the disease from unaffected parents. This is a hallmark of recessive inheritance. The pedigree is also most consistent with the allele being X-linked for the following reasons: only males are affected; both affected progeny could have inherited the allele through their maternal parents who were either the daughter or grand-daughter of the original affected male; and you are told that the disease is rare so if it is not X-linked, both fathers of affected individuals that have married into this pedigree would also have to be carriers.

Pedigree 2: autosomal recessive

As in pedigree 1, the trait appears to be recessive. (This pedigree also indicates why it is more likely to see rare recessive traits in the progeny of consanguineous unions.) In this pedigree, it is more likely that the trait is autosomal. If the trait was X-linked, the affected son in the fourth generation would have inherited the trait from his mother, yet his mother could not have inherited an X-linked trait from her unaffected father. Again, you are told that the disease is rare so it is less likely that this affected son's grandmother was also a carrier.

b. Child of A and B:

Since this is an X-linked recessive trait, you must calculate B's probability of being a carrier; if she is, you must figure the probability of her passing that allele to her child and multiply that by the probability that her child will be male.

B's grandmother must be a carrier. Thus, B's mother has a 50% chance of being a carrier and a 50% chance that if she is, she passed the allele to B. If B is a carrier, her children have a 50% chance of inheriting the allele and, if male, being affected. Therefore, A and B have a $1/2 \times 1/2 \times 1/2 \times 1/2 = 1/16$ chance of having an affected child.

Child of C and D:

This trait is inherited as an autosomal recessive, therefore both C and D must be carriers for their child to have a 25% chance of being affected. Although D's probability can be calculated as $1/3$ since her mother had a $2/3$ chance of being heterozygous, we do not know C's genotype. Since the trait is rare, we can assume that C has 0% chance of being a carrier and 0% chance of passing the allele to his child.

5 Recombination of Genes

1. You are told that the two genes assort independently. Therefore,

From the first parent: $^1/_4$ the gametes will be A ; B
$^1/_4$ the gametes will be A ; b
$^1/_4$ the gametes will be a ; B
$^1/_4$ the gametes will be a ; b

From the second parent: $^1/_2$ the gametes will be A ; b
$^1/_2$ the gametes will be a ; b

and the progeny will be: $^1/_8$ A/A ; B/b
$^1/_8$ A/A ; b/b
$^1/_4$ A/a ; B/b
$^1/_4$ A/a ; b/b
$^1/_8$ a/a ; B/b
$^1/_8$ a/a ; b/b

2. **a., b.** You are told that black is dominant to brown and intense is dominant to dilute. Thus, the brown and the dilute (recessive) traits must be homozygous to be expressed. Neither the black nor intense traits breed true, so these must both be heterozygous.

Parents: dilute ; Black × Intense ; brown
d/d ; B/b D/d ; b/b

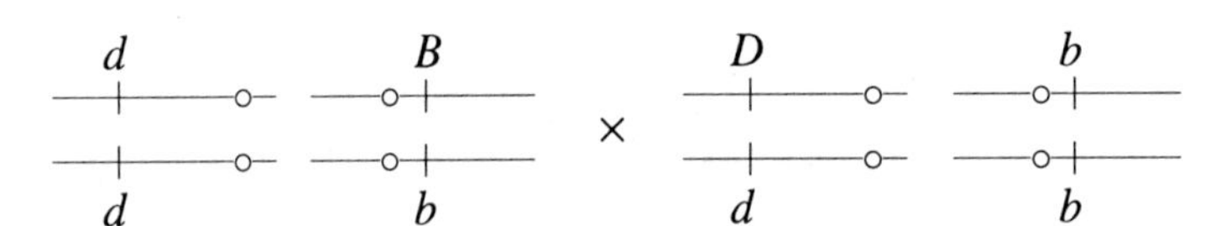

Progeny:		
	dilute ; Black	d/d ; B/b
	dilute ; brown	d/d ; b/b
	Intense ; Black	D/d ; B/b
	Intense ; brown	D/d ; b/b

3. **a.** 9 genotypes (just count)

b.

9:	3:	3:	1:
1 R/R ; Y/Y	1 R/R ; y/y	1 r/r ; Y/Y	1 r/r ; y/y
2 R/R ; Y/y	2 R/r ; y/y	2 r/r ; Y/y	
4 R/r ; Y/y			
2 R/r ; Y/Y			

c. The number of different genotypes is 3^n, where n = number of genes. For simple dominant/recessive relationships, the number of different phenotypes is 2^n, where n = number of genes.

d. A round, yellow plant's genotype can be deduced either through a self-cross or testcross.

4. You perform the following cross and are told that the two genes are 10 m.u. apart.

$$A\ B/a\ b \times a\ b/a\ b$$

Among their progeny, 10% should be recombinant ($A\ b/a\ b$ and $a\ B/a\ b$) and 90% should be parental ($A\ B/a\ b$ and $a\ b/a\ b$). Therefore, $A\ B/a\ b$ should represent $^1/_2$ of the parentals or 45%.

5. P $A\ d\ /\ A\ d$ × $a\ D\ /\ a\ D$

F_1 $A\ d\ /\ a\ D$

F_2 1 $A\ d\ /\ A\ d$ Phenotype: A d
2 $A\ d\ /\ a\ D$ Phenotype: A D
1 $a\ D\ /\ a\ D$ Phenotype: a D

6. Since only parental types are recovered, the two genes must be tightly linked and recombination must be very rare. Knowing how many progeny were looked at would give an indication of how close the genes are.

7. The problem states that a female that is *A/a. B/b* is testcrossed. If the genes are unlinked, they should assort independently and the four progeny classes should be present in roughly equal proportions. This is clearly not the case. The *A/a . B/b* and *a/a . b/b* classes (the parentals) are much more common than the *A/a . b/b* and *a/a . B/b* classes (the recombinants). The two genes are on the same chromosome and are 10 map units apart.

RF = 100% × (46 + 54)/1000 = 10%

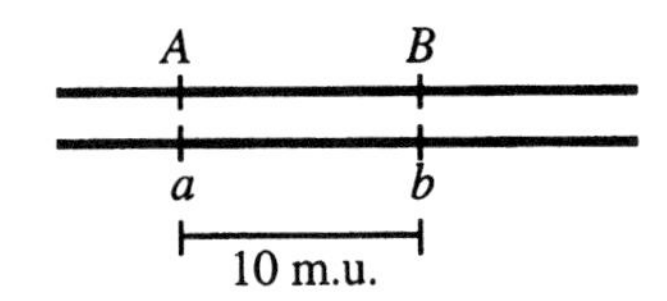

8. The cross is *A/A . b/b* × *a/a . B/B*. The F_1 would be *A/a . B/b*.

a. If the genes are unlinked, all four progeny classes from the testcross (including *a/a ; b/b*) would equal 25%.

b. With completely linked genes, the F_1 would produce only *A b* and *a B* gametes. Thus, there would be a 0% chance of having *a b/a b* progeny from a testcross of this F_1.

c. If the two genes are linked and 12 map units apart, 12% of the testcross progeny should be recombinants. Since the F_1 is *A b/a B*, *a b* is one of the recombinant classes (*A B* being the other) and should equal $^1/_2$ of the total recombinants or 6%.

d. 12% (see part c)

9. To answer this question, you must realize that (1) one chiasma involves two of the four chromatids of the homologous pair, so if 16% of the meioses have one chiasma, it will lead to 8% recombinants observed in the progeny (one half of the chromosomes of such a meiosis are still parental); and (2) half of the recombinants will be *B r*, so the correct answer is 4% (or b).

10.

a.	*C/c ; S/s* × *C/c ;*	*S/s*	There are 3 short:1 long, and 3 dark:1 albino.
b.	*C/C ; S/s* × *C/C ;*	*s/s*	There are no albino, and there are 1 long: 1 short.
c.	*C/c ; S/S* × *c/c ;*	*S/S*	There are no long, and there are 1 dark: 1 albino.
d.	*c/c ; S/s* × *c/c ;*	*S/s*	All are albino, and there are 3 short:1 long.
e.	*C/c ; s/s* × *C/c ;*	*s/s*	All are long, and there are 3 dark:1 albino.
f.	*C/C ; S/s* × *C/C ;*	*S/s*	There are no albino, and there are 3 short: 1 long.
g.	*C/c ; S/s* × *C/c ;*	*s/s*	There are 3 dark:1 albino, and 1 short:1 long.

11. Cross 2 indicates that purple (*G*) is dominant to green (*g*), and cross 1 indicates cut (*P*) is dominant to potato (*p*).

Cross 1:	*G/g ; P/p* × *g/g ; P/p*	There are 3 cut:1 potato, and 1 purple:1 green.
Cross 2:	G/g ; *P/p* × *G/g ; p/p*	There are 3 purple:1 green, and 1 cut:1 potato.
Cross 3:	*G/G ; P/p* × *g/g ; P/p*	There are no green, and there are 3 cut:1 potato.
Cross 4:	*G/g ; P/P* × *g/g ; p/p*	There are no potato, and there are 1 purple: 1 green.
Cross 5:	*G/g ; p/p* × *g/g ; P/p*	There are 1 cut:1 potato, and there are 1 purple:1 green.

12. a. Since each gene assorts independently, each probability should be considered separately and then multiplied together for the answer.

For (1) $^1/_2$ will be A, $^3/_4$ will be B, $^1/_2$ will be C, $^3/_4$ will be D, and $^1/_2$ will be E.

$$^1/_2 \times {}^3/_4 \times {}^1/_2 \times {}^3/_4 \times {}^1/_2 = {}^9/_{128}$$

For (2) $^1/_2$ will be a, $^3/_4$ will be B, $^1/_2$ will be c, $^3/_4$ will be D, and $^1/_2$ will be e.

$$^1/_2 \times {}^3/_4 \times {}^1/_2 \times {}^3/_4 \times {}^1/_2 = {}^9/_{128}$$

For (3) it is the sum of (1) and (2) = $^9/_{128} + {}^9/_{128} = {}^9/_{64}$

For (4) it is 1 – (part 3) = $1 - {}^9/_{64} = {}^{55}/_{64}$

b. For (1) $^1/_2$ will be *A/a*, $^1/_2$ will be *B/b*, $^1/_2$ will be *C/c*, $^1/_2$ will be *D/d*, and $^1/_2$ will be *E/e*.

$$^1/_2 \times {}^1/_2 \times {}^1/_2 \times {}^1/_2 \times {}^1/_2 = {}^1/_{32}$$

For (2) $^1/_2$ will be *a/a*, $^1/_2$ will be *B/b*, $^1/_2$ will be *c/c*, $^1/_2$ will be *D/d*, and $^1/_2$ will be *e/e*.

$$^1/_2 \times {}^1/_2 \times {}^1/_2 \times {}^1/_2 \times {}^1/_2 = {}^1/_{32}$$

For (3) it is the sum of (1) and (2) = $^1/_{16}$

For (4) it is 1 – (part 3) = $1 - {}^1/_{16} = {}^{15}/_{16}$

13. Assume there is no linkage. (This is your hypothesis. If it can be rejected, the genes are linked.) The expected values would be that genotypes occur with equal frequency. There are four genotypes in each case ($n = 4$) so there are 3 degrees of freedom.

$$\chi^2 = \Sigma\,(\text{observed} - \text{expected})^2/\text{expected}$$

Cross 1: $\chi^2 = [(310 - 300)^2 + (315 - 300)^2 + (287 - 300)^2 + (288 - 300)^2]/300$
$= 2.1266$; $p > 0.50$, nonsignificant; hypothesis cannot be rejected

Cross 2: $\chi^2 = [(36 - 30)^2 + (38 - 30)^2 + (23 - 30)^2 + (23 - 30)^2]/300$
$= 6.6$; $p > 0.10$, nonsignificant; hypothesis cannot be rejected

Cross 3: $\chi^2 = [(360 - 300)^2 + (380 - 300)^2 + (230 - 300)^2 + (230 - 300)^2]/300$
$= 66.0$; $p < 0.005$, significant; hypothesis must be rejected

Cross 4: $\chi^2 = [(74 - 60)^2 + (72 - 60)^2 + (50 - 60)^2 + (44 - 60)^2]/300$
$= 11.60$; $p < 0.01$, significant; hypothesis must be rejected

14. a. Note that only males are affected by both disorders. This suggests that both are X-linked recessive disorders. Using *p* for protan and *P* for nonprotan and *d* for deutan and *D* for nondeutan, the inferred genotypes are listed on the pedigree below. The Y chromosome is shown, but the X is represented by the alleles carried.

b. Individual II-2 must have inherited both disorders in the trans-configuration (on separate chromosomes). Therefore, individual III-2 inherited both traits as the result of recombination (crossing-over) between his mother's X chromosomes.

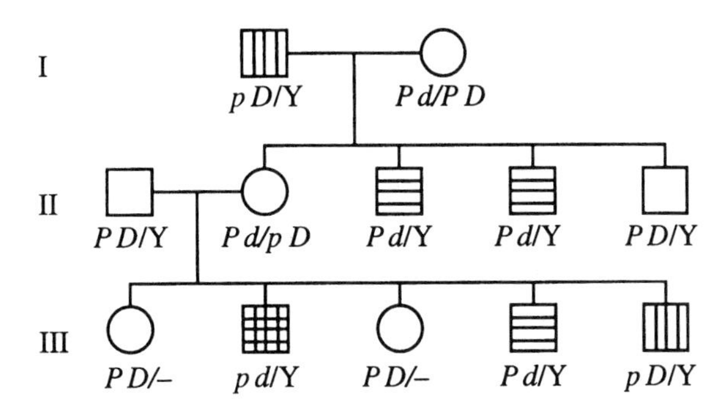

c. Since both genes are X-linked, this represents crossing-over. The progeny size is too small to give a reliable estimate of recombination.

15. ***Unpacking the Problem***

a. There is no correct drawing; any will do. Pollen from the tassels is placed on the silks of the females. The seeds are the F_1 corn kernels.

b. The +'s all look the same because they signify wild type for each gene. The information is given in a specific order, which prevents confusion, at least initially. However, as you work the problem, which may require you to reorder the genes, errors can creep into your work if you do not make sure that you reorder the genes for each genotype in exactly the same way. You may find it easier to write the complete genotype, p^+ instead of +, to avoid confusion.

c. The phenotype is purple leaves and brown midriff to seeds. In other words, the two colors refer to different parts of the organism.

d. There is no significance in the original sequence of the data.

e. A tester is a homozygous recessive for all genes being studied. It is used so that the meiotic products in the organism being tested can be seen directly in the phenotype of the progeny.

f. The progeny phenotypes allow you to infer the genotypes of the plants. For example, *gre* stands for "green," the phenotype of $p^+/–$; *sen* stands for "virus-sensitive," the phenotype of $v^+/–$; and *pla* stands for "plain seed," the phenotype of $b^+/–$. In this testcross, all progeny have at least one recessive allele so the "gre sen pla" progeny are actually $p^+/p\ .\ v^+/v\ .\ b^+/b$.

g. *Gametes* refers to the gametes of the two pure-breeding parents. *F_1 gametes* refers to the gametes produced by the completely heterozygous F_1 progeny. They indicate whether crossing-over and independent assortment have occurred. In this case, because there is either independent assortment or crossing-over, or both, the data indicate that the three genes are not so tightly linked that zero recombination occurred.

h. The main focus is meiosis occurring in the F_1 parent.

i. The gametes from the tester are not shown because they contribute nothing to the phenotypic differences seen in the progeny.

j. Eight phenotypic classes are expected for three autosomal genes, whether or not they are linked, when all three genes have simple dominant-recessive relationships among their alleles. The general formula for the number of expected phenotypes is 2^n , where n is the number of genes being studied.

k. If the three genes were on separate chromosomes, the expectation is a 1:1:1:1:1:1:1:1 ratio.

l. The four classes of data correspond to the parentals (largest), two groups of single crossovers (intermediate), and double crossovers (smallest).

m. By comparing the parentals with the double crossovers, gene order can be determined. The gene in the middle "flips" with respect to the two flanking genes in the double-crossover progeny. In this case, one parental is +++ and one double crossover is *p*++. This indicates that the gene for leaf color (*p*) is in the middle.

n. If only two of the three genes are linked, the data can still be grouped, but the grouping will differ from that mentioned in (l) above. In this situation, the unlinked gene will show independent assortment with the two linked genes. There will be one class composed of four phenotypes in approximately equal frequency, which combined will total more than half the progeny. A second class will be composed of four phenotypes in approximately equal frequency, and the combined total will be less than half the progeny. For example, if the cross were *a b*/+ + ; *c*/+ × *a b*/*a b* ; *c*/*c*, then the parental class (more frequent class) would have four components: *a b c*, *a b* +, + + *c*, and + + +. The recombinant class would be *a* + *c*, *a* + +, + *b c*, and + *b* +.

o. *Point* refers to locus. The usage does not imply linkage, but rather a testing for possible linkage. A four-point testcross would look like the following: *a*/+ . *b*/+ . *c*/+ . *d*/+ × *a*/*a* . *b*/*b* . *c*/*c* . *d*/*d*.

p. A *recombinant* refers to an individual who has alleles inherited from two different grandparents, both of whom were the parents of the individual's heterozygous parent. Another way to think about this term is that in the recombinant individual's heterozygous parent, recombination took place among the genes that were inherited from his or her parents. In this case, the recombination took place in the F_1 and the recombinants are among the F_2 progeny.

q. The "recombinant for" columns refer to specific gene pairs and progeny that exhibit recombination between those gene pairs.

r. There are three "recombinant for" columns because three genes can be grouped in three different gene pairs.

s. *R* refers to recombinant progeny, and they are determined by reference back to the parents of their heterozygous parent.

t. Column totals indicate the number of progeny that experience crossing-over between the specific gene pairs. They are used to calculate map units between the two genes.

u. The diagnostic test for linkage is a recombination frequency of less than 50%.

v. A map unit represents 1% crossing-over and is the same as a centimorgan.

w. In the tester, recombination cannot be detected in the gamete contribution to the progeny because the tester is homozygous. The F_1 individuals have genotypes fixed by their parents' homozygous state and, again, recombination cannot be detected in them, simply because their parents were homozygous.

x. Interference I = 1 – coefficient of coincidence = 1 – (observed double crossovers/expected double crossovers). The expected double crossovers are equal to p(frequency of crossing-over in the first region, in this case between v and p) × p(frequency of crossing-over in the second region, between p and b) × number of progeny. The probability of crossing-over is equal to map units converted back to percentage.

y. If the three genes are not all linked, then interference cannot be calculated.

z. A great deal of work is required to obtain 10,000 progeny in corn because each seed on a cob represents one progeny. Each cob may contain as many as 200 seeds. While seed characteristics can be assessed at the cob stage, for other characteristics, each seed must separately be planted and assessed after germination and growth. The bookkeeping task is also enormous.

Solution to the Problem

a. The three genes are linked.

b. Comparing the parentals (most frequent) with the double crossovers (least frequent), the gene order is $v\ p\ b$. There were 2200 recombinants between v and p, and 1500 between p and b. The general formula for map units is

m.u. = 100%(number of recombinants)/total number of progeny

Therefore, the map units between v and p = 100%(2200)/10,000 = 22 m.u., and the map units between p and b = 100%(1500)/10,000 = 15 m.u.

The map is

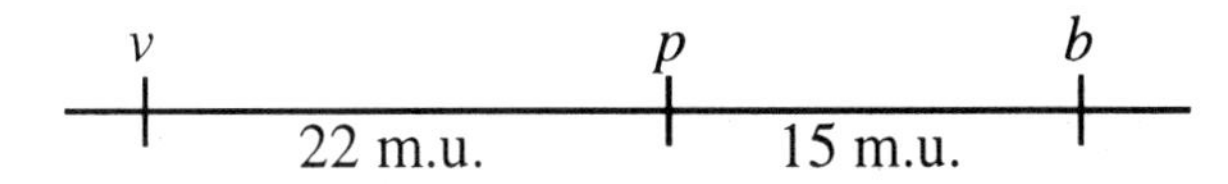

c. I = 1 – observed double crossovers/expected double crossovers
= 1 – 132/(0.22)(0.15)(10,000)
= 1 – 0.4 = 0.6

16. a. By comparing the two most frequent classes (parentals: $an\ br^+\ f^+$, $an^+ br\ f$) to the least frequent classes (DCO: $an^+\ br\ f^+$, $an\ br^+\ f$), the gene order can be determined. The gene in the middle switches with respect to the other two (the order is *an f br*). Now the crosses can be written fully

P	$an\ f^+\ br^+/an\ f^+\ br^+$	×	$an^+\ f\ br/an^+\ f\ br$
F_1	$an\ f^+\ br^+/an^+\ f\ br$	×	$an\ f\ br/an\ f\ br$

F_2	355	$an\ f^+\ br^+/an\ f\ br$	Parental
	339	$an^+\ f\ br/an\ f\ br$	Parental
	88	$an^+\ f^+\ br^+/an\ f\ br$	CO *an–f*
	55	$an\ f\ br/an\ f\ br$	CO *an–f*
	21	$an^+\ f\ br^+/an\ f\ br$	CO *f–br*
	17	$an\ f^+\ br/an\ f\ br$	CO *f–br*
	2	$an^+\ f^+\ br/an\ f\ br$	DCO
	2	$an\ f\ br^+/an\ f\ br$	DCO

b. *an–f*: 100%(88 + 55 + 2 + 2)/879 = 16.72 m.u.
f–br: 100%(21 + 17 + 2 + 2)/879 = 4.78 m.u.

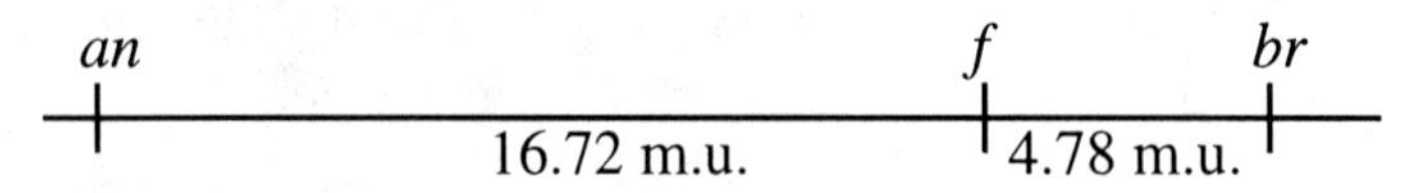

c. Interference = 1 – (observed DCO/expected DCO)
= 1 – 4/(0.1672)(0.0478)(879)
= 1 – 0.569 = 0.431

17. The data given for each of the three-point testcrosses can be used to determine the gene order by realizing that the rarest recombinant classes are the result of double crossover events. By comparing these chromosomes to the "parental" types, the alleles that have switched represent the gene in the middle.

For example, in (1), the most common phenotypes (+ + + and a b c) represent the parental allele combinations. Comparing these to the rarest phenotypes of this data set (+ b c and a + +) indicates that the *a* gene is recombinant and must be in the middle. The gene order is *b a c*.

For (2), + b c and a + + (the parentals) should be compared to + + + and a b c (the rarest recombinants) to indicate that the *a* gene is in the middle. The gene order is *b a c*.

For (3), compare + b + and a + c with a b + and + + c, which gives the gene order *b a c*.

For (4), compare + + c and a b + with + + + and a b c, which gives the gene order *a c b*.

For (5), compare + + + and a b c with + + c and a b +, which gives the gene order *a c b*.

18. a. The first F_1 is *L H/l h* and the second is *l H/L h*. For progeny that are *l h/l h*, they have received a "parental" chromosome from the first F_1 and a "recombinant" chromosome from the second F_1. The genes are 16% apart so the chance of a parental chromosome is $^1/_2$(100 – 16%) = 42% and the chance of a recombinant chromosome is $^1/_2$(16%) = 8%.

The chance of both events = 42% × 8% = 3.36%

b. To obtain *Lh/l h* progeny, either a parental chromosome from each parent was inherited *or* a recombinant chromosome from each parent was inherited. The total probability will therefore be

$$(42\% \times 42\%) + (8\% \times 8\%) = (17.6\% + 0.6\%) = 18.2\%$$

19. Since there is no branch migration, the heteroduplex that occurs is only the result of strand invasion. All heteroduplexes are repaired, but there is a bias to repair the mismatch to the *A* allele over the *a* allele (80% to *A* but only 20% to *a*).

Strand invasion by *A* would result in a transient heteroduplex (mismatch) that would be repaired to *A* 80% of the time and result in an aberrant 6:2 ratio (*A*:*a*). In the other 20% of these cases, the *A* would be converted to an *a*, resulting in the normally expected 4:4 ratio. Strand invasion by *a* would result in a transient heteroduplex that would be repaired to *a* 20% of the time and result in an aberrant 2:6 ratio (*A*:*a*). In the other 80% of these cases, the *a* would be converted to an *A*, resulting in the normally expected 4:4 ratio. Overall then, 80% of the aberrant asci will show a 6:2 ratio and 20% will show a 2:6 ratio. Since all heteroduplexes are repaired, no 5:3 or 3:5 aberrant ratios will be observed.

20.

(1) Impossible:	The alleles *A* and *a* should be on homologous chromosomes as should the alleles *B* and *b*.
(2) Meiosis II:	Sister chromatids are separating, and there is only one copy of each chromosome.
(3) Meiosis II:	Same as (2).
(4) Meiosis II:	Same as (2).
(5) Mitosis:	Sister chromatids are separating, and there are two copies of each chromosome.
(6) Impossible:	Sister chromatids are nonidentical for all chromosomes.
(7) Impossible:	There are 4 copies of each chromosome.
(8) Impossible:	Same as (7).
(9) Impossible:	Same as (7).
(10) Meiosis I:	Homologous chromosomes are separating.
(11) Impossible:	All four chromatids of each homologous chromosome have the same allele.
(12) Impossible:	Same as (1).

21. The rare prototrophic colonies could be the result of intragenic recombination between the two mutations of the *hist-1* gene. Since both mutations *never* revert, assume that both are the result of small deletions. Reciprocal recombination between these will result in one wild-type allele and one doubly mutant allele (which would not appear different than either single mutation). Only the wild-type alleles would be observed, and these would represent one half of the total recombinants.

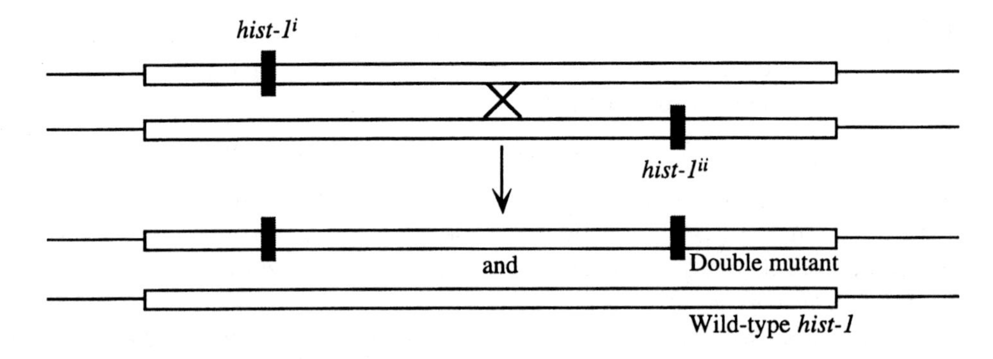

22. Dear Monk Mendel:

I have recently read your most engrossing manuscript detailing the results of your most wise experiments with garden peas. I salute both your curiosity and your ingenuity in conducting said experiments, thereby opening up for scientific exploration an entire new area of our Maker's universe. Dear Sir, your findings are extraordinary!

While I do not pretend to compare myself to you in any fashion, I beg to bring to your attention certain findings I have made with the aid of that most fascinating and revealing instrument, the microscope. I have been turning my attention to the smallest of worlds with an instrument that I myself have built, and I have noticed some structures that may parallel in behavior the factors that you have postulated in the pea.

I have worked with grasshoppers, however, not your garden peas. Although you are a man of the cloth, you are also a man of science, and I pray that you will not be offended when I state that I have specifically studied the reproductive organs of male grasshoppers. Indeed, I did not limit myself to studying the organs themselves; instead, I also studied the smaller units that make up the male organs and have beheld structures most amazing within them.

These structures are contained within numerous small bags within the male organs. Each bag has a number of these structures, which are long and threadlike at some times and short and compact at other times. They come together in the middle of a bag, and then they appear to divide equally. Shortly thereafter, the bag itself divides, and what looks like half of the threadlike structures goes into each new bag. Could it be, Sir, that these threadlike structures are the very same as your factors? I know, of course, that garden peas do not have male organs in the same way that grasshoppers do, but it seems to me that you found it necessary to emasculate the garden peas in order to do some crosses, so I do not think it too far-fetched to postulate a similarity between grasshoppers and garden peas in this respect.

Pray, Sir, do not laugh at me and dismiss my thoughts on this subject even though I have neither your excellent training nor your astounding wisdom in the Sciences. I remain your humble servant to eternity!

23. Two unlinked genes; "no dots" (*D*) is dominant to "dots" (*d*) and "vertical bar" (*H*) is dominant to "horizontal bar" (*h*).

dots, Vertical bar × No dots, horizontal bar
$d/d\,;\,H/H$ × $D/D\,;\,h/h$

↓

All: No dots, Vertical bar × dots, horizontal bar
$D/d\,;\,H/h$ × $d/d\,;\,h/h$

↓

$^1/_4$ dots, horizontal bar
$d/d\,;\,h/h$

$^1/_4$ No dots, horizontal bar
$D/d\,;\,h/h$

$^1/_4$ dots, Vertical bar
$d/d\,;\,H/h$

$^1/_4$ No dots, Vertical bar
$D/d\,;\,H/h$

24. Two unlinked genes; "no dots" (*d*) is recessive to "dots" (d^+) and "no lines" (*l*) is recessive to "lines" (l^+).

no dots, Lines × Dots, no lines
$d/d\,;\,l^+/l^+$ × $d^+/d^+\,;\,l/l$

↓

All: Dots, Lines
$d^+/d\,;\,l^+/l$

↓

Self: $^9/_{16}$ Dots, Lines
$d^+/-\,;\,l^+/-$

$^3/_{16}$ Dots; no lines
$d^+/-\,;\,l/l$

$^3/_{16}$ no dots, Lines
$d/d\,;\,l^+/-$

$^1/_{16}$ no dots, no lines
$d/d\,;\,l/l$

25. Two linked genes, 20 m.u. apart; "no dots" (d) is recessive to "dots" (D) and "no line" (l) is recessive to "line" (L).

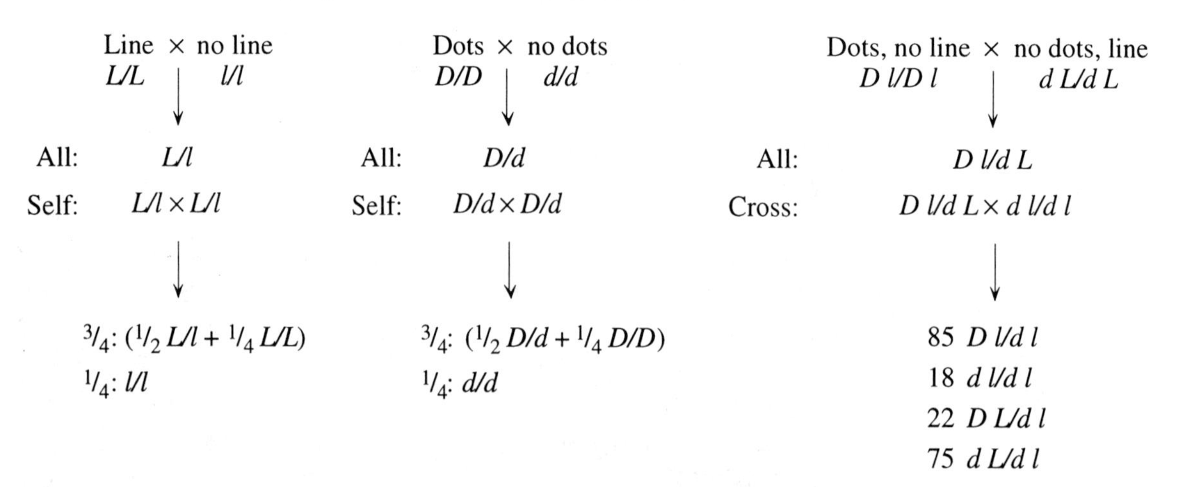

26. Two X-linked genes, 10 m.u. apart; "white" (b) is recessive to "black" (b^+) and "wavy tail" (s) is recessive to "straight tail" (s^+).

P $\quad b^+ s^+/b^+ s^+ \times b\, s/Y$

$\downarrow$

F_1 $\quad b^+ s^+/b\, s$ and $b^+ s^+/Y$

$\downarrow$

F_2 $\quad$ Females: Although all phenotypically $b^+ s^+$, genotypically they are

45% $b^+ s^+/b^+ s^+$
45% $b\, s/b^+ s^+$
5% $b\, s^+/b^+ s^+$
5% $b^+ s/b^+ s^+$

Males: 45% $b^+ s^+/Y$
45% $b\, s/Y$
5% $b\, s^+/Y$
5% $b^+ s/Y$

27. Two linked genes, 30 m.u. apart. (Remember, fungi are haploid.)

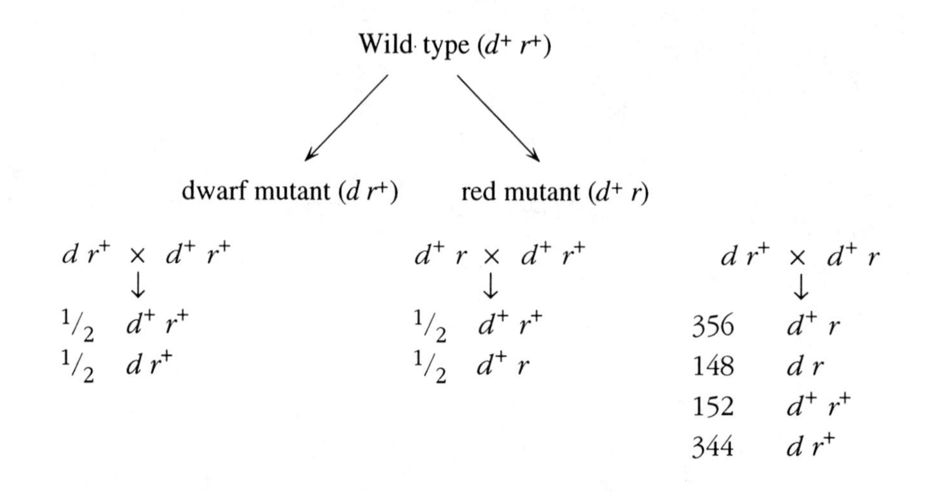

28. a. Long (s^+) is dominant to short (s), since the second cross has short progeny from long parents.

Large (v^+) is dominant to vestigial (v), since the third cross has vestigial progeny from large parents.

b. Since short and vestigial are recessive, only flies homozygous for the respective alleles will express these traits.

Cross 1:	v^+/v ; s^+/s^+ × v/v ; s/s	Long parent is true-breeding, large is not.
Cross 2:	v^+/v ; s^+/s × v/v ; s^+/s	Large parent not true-breeding; 3 long:1 short.
Cross 3:	v^+/v ; s^+/s × v^+/v ; s/s	Long parent not true-breeding; 3 large:1 vestigial.
Cross 4:	v^+/v ; s/s × v/v ; s^+/s	Neither large nor long parents are true-breeding.

29. a., b. It is likely that the observed abnormalities are the result of mitotic recombination. A crossover between the genes of interest and the centromere in this case will lead to "twin spots" (the adjacent patches of stubby and ebony body observed). On the other hand, a crossover that occurs between the two genes will lead to "single spots" (the solitary patches of ebony). The position of the two genes with respect to the centromere determines the phenotype of the single spot. When recombination occurs in the region between the genes, the gene more distal to the centromere becomes homozygous while the more proximal gene remains heterozygous. In this problem, since the single patches are ebony, *e* is more distal than *s*. The other product of this event will appear normal and will not be detected among the other "normal" cells.

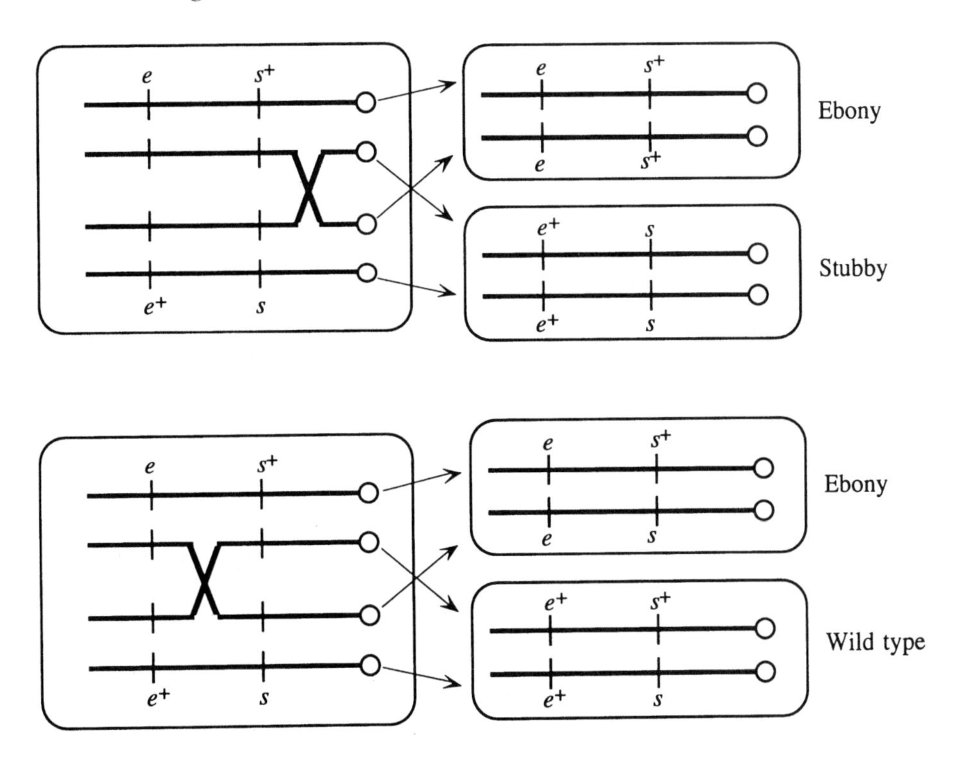

30. a., b., c. Since all eight allelic combinations are equally likely in the gametes, it can be inferred that the three genes are on separate chromosomes. The following figures use chromosome size and centromere placement to distinguish the three nonhomologous chromosomes.

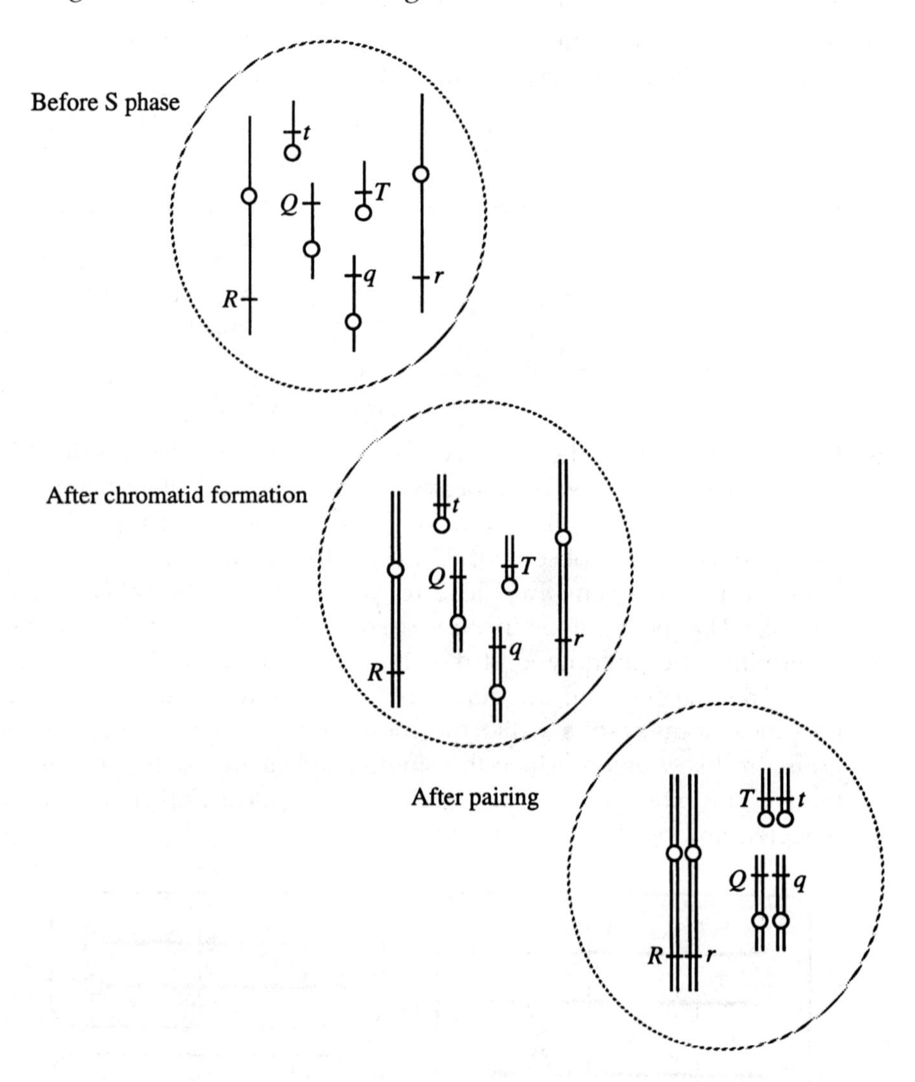

d., e. Independent assortment of nonhomologous chromosomes will give eight possible allelic combinations. This is graphically represented by showing the various alignments the separate chromosomes may take during meiosis. The segregation of homologous chromosomes during anaphase 1 and then the splitting of sister chromatids during anaphase II is schematically indicated. Although it is very likely that crossing-over will take place during prophase I, it will not affect the genotypic ratios of the gametes and thus is ignored.

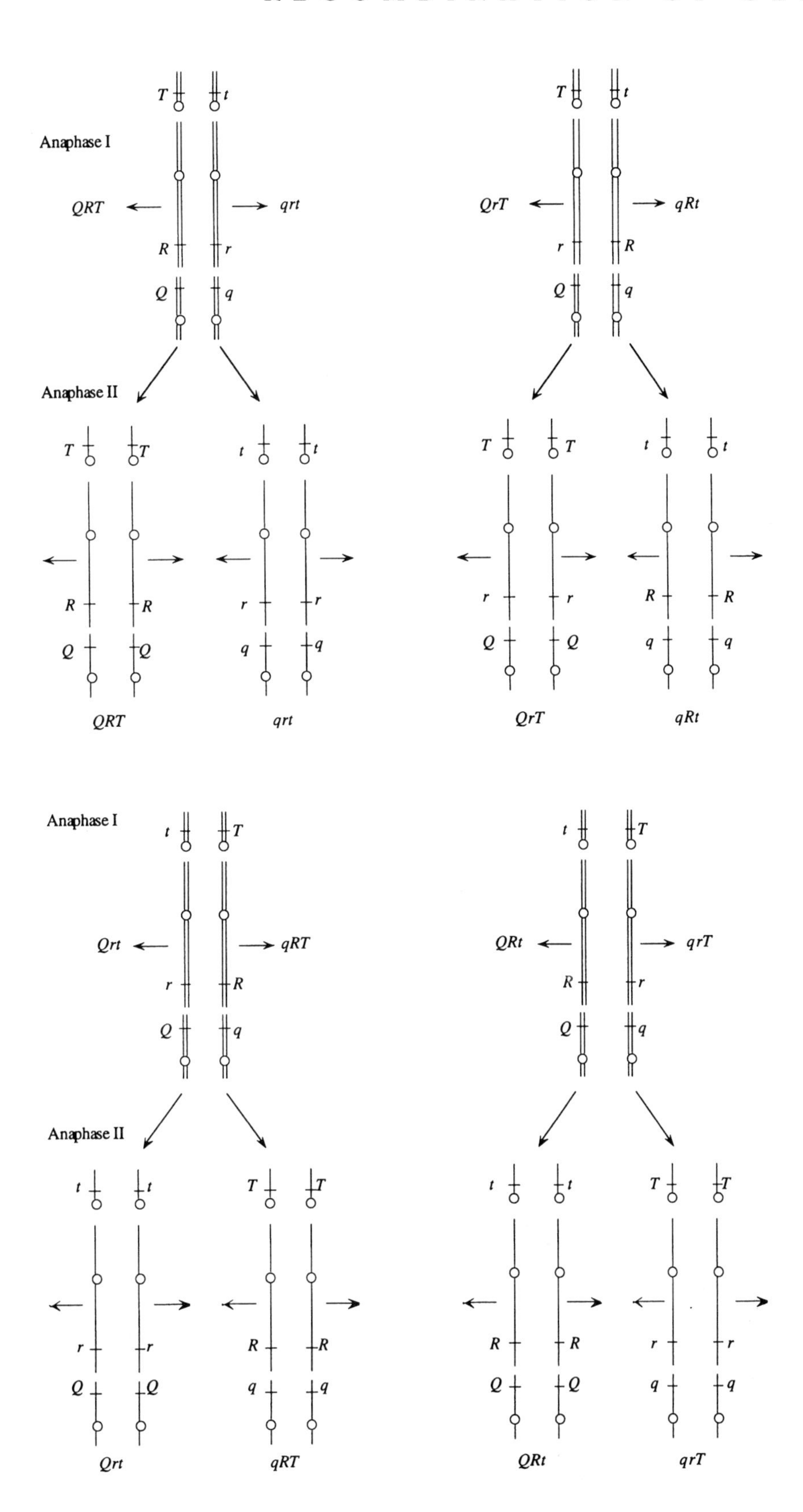
Anaphase I
QRT
qrt
QrT
qRt
Anaphase II
QRT
qrt
QrT
qRt
Anaphase I
Qrt
qRT
QRt
qrT
Anaphase II
Qrt
qRT
QRt
qrT

31. a. From the data, it can be concluded that the genes for flower color and plant height are linked (on the same chromosome). The position of the gene for leaf width cannot be determined from this data since all progeny express the dominant trait. In the tall, red parent, the alleles for tall and white are in the cis configuration (on the same chromosome) as are the alleles for short and red. Thus, the progeny that are tall, white or short, red represent the parental chromosomes and the tall, red or short, white are recombinants. The two genes are 100%(21+19)/total = 4 map units apart.

Tall, Red, Wide × short, white, narrow

$$\frac{T \quad r}{t \quad R} \cdot \frac{W}{W} \times \frac{t \quad r}{t \quad r} \cdot \frac{w}{w}$$

b. The chance of obtaining short, white, wide progeny = $p(t\ r)^2 = (^{4\%}/_{2})^2 =$ 0.04%.

6 Gene Interaction

1. Complementation is the cooperation of the products of two or more genes to produce a nonmutant phenotype, while recombination is the exchange of DNA segments between chromosomes. Recombination can occur both within and between genes, while complementation occurs between gene products.

2. You are told that the cross of two erminette fowls results in 22 erminette, 14 black, and 12 pure white. Two facts are important: (1) the parents consist of only one phenotype, yet the offspring have three phenotypes, and (2) the progeny appear in an approximate ratio of 1:2:1. These facts should tell you immediately that you are dealing with a heterozygous × heterozygous cross involving one gene and that the erminette phenotype must be the heterozygous phenotype.

   When the heterozygote shows a different phenotype from either of the two homozygotes, the heterozygous phenotype results from incomplete dominance or codominance. Because two of the three phenotypes contain black, either fully or in an occasional feather, you might classify erminette as an instance of incomplete dominance because it is intermediate between fully black and fully white. Alternatively, because erminette has both black and white feathers, you might classify the phenotype as codominant. Your decision will rest on whether you look at the whole animal (incomplete dominance) or at individual feathers (codominance). This is yet another instance where what you conclude is determined by how you observe.

   To test the hypothesis that the erminette phenotype is a heterozygous phenotype, you could cross an erminette with either or both of the homozygotes. You should observe a 1:1 ratio in the progeny of both crosses.

3. From the cross $c^+/c^{ch} \times c^{ch}/c^h$ the progeny are

$^1/_4$	c^+/c^{ch}	Full color
$^1/_4$	c^+/c^h	Full color
$^1/_4$	c^{ch}/c^{ch}	Chinchilla
$^1/_4$	c^{ch}/c^h	Chinchilla

Thus, 50% of the progeny will be chinchilla.

4. **a.** The data indicate that there is a single gene with multiple alleles. The order of dominance is

Black > sepia > cream > albino

Cross	Parents	Progeny	Conclusion
Cross 1:	$b/a \times b/a$	3 $b/–$:1 a/a	Black is dominant to albino.
Cross 2:	$b/s \times a/a$	1 b/a:1 s/a	Black is dominant to sepia; sepia is dominant to albino.
Cross 3:	$c/a \times c/a$	3 $c/–$:1 a/a	Cream is dominant to albino.
Cross 4:	$s/a \times c/a$	1 c/a:2 $s/–$:1 a/a	Sepia is dominant to cream.
Cross 5:	$b/c \times a/a$	1 b/a:1 c/a	Black is dominant to cream.
Cross 6:	$b/s \times c/–$	1 $b/–$:1 $s/–$	"–" can be *c* or *a*.
Cross 7:	$b/s \times s/–$	1 b/s:1 $s/–$	"–" can be *s*, *c*, or *a*.
Cross 8:	$b/c \times s/c$	1 s/c:2 $b/–$:1 c/c	"–" can be *s* or *c*.
Cross 9:	$s/c \times s/c$	3 $s/–$:1 c/c	"–" can be *s* or *c*.
Cross 10:	$c/a \times a/a$	1 c/a:1 a/a	

b. The progeny of the cross $b/s \times b/c$ will be $^3/_4$ black ($^1/_4$ b/b, $^1/_4$ b/c, $^1/_4$ b/s):$^1/_4$ sepia (s/c).

5. Both codominance (=) and classical dominance (>) are present in the multiple allelic series for blood type: $A = B$, $A > O$, $B > O$.

Parents' phenotype	Parents' possible genotypes	Parents' possible children
a. AB × O	$A/B \times O/O$	A/O, B/O
b. A × O	A/A or $A/O \times O/O$	A/O, O/O
c. A × AB	A/A or $A/O \times A/B$	A/A, A/B, A/O, B/O
d. O × O	$O/O \times O/O$	O/O

The possible genotypes of the children are:

Phenotype	Possible genotypes
O	O/O
A	A/A, A/O
B	B/B, B/O
AB	A/B

Using the assumption that each set of parents had one child, the following combinations are the only ones that will work as a solution:

Parents	Child
a. AB × O	B
b. A × O	A
c. A × AB	AB
d. O × O	O

6. **a.** The sex ratio is expected to be 1:1.

b. The female parent was heterozygous for an X-linked recessive lethal allele. This would result in 50% fewer males than females.

c. Half of the female progeny should be heterozygous for the lethal allele and half should be homozygous for the nonlethal allele. Individually mate the F_1 females and determine the sex ratio of their progeny.

7. Note that the F_2 are in a 9:6:1 ratio. This suggests a dihybrid cross in which *A/– ; b/b* has the same appearance as *a/a ; B/–*. Let the disc phenotype be the result of *A/– ; B/–* and the long phenotype be the result of *a/a ; b/b*. The crosses are

P	*A/A ; B/B* (disc) × *a/a ; b/b* (long)	
F_1	*A/a ; B/b* (disc)	
F_2	9	*A/– ; B/–* (disc)
	3	*a/a ; B/–* (sphere)
	3	*A/– ; b/b* (sphere)
	1	*a/a ; b/b* (long)

8. The suggestion from the data is that the two albino lines had mutations in two different genes. When the extracts from the two lines were placed in the same test tube, they were capable of producing color because the gene product of one line was capable of compensating for the absence of a gene product from the second line.

a. The most obvious control is to cross the two pure-breeding lines. The cross would be *A/A ; b/b* × *a/a ; B/B*. The progeny will be *A/a ; B/b*, and all should be reddish purple.

b. The most likely explanation is that the red pigment is produced by the action of at least two different gene products.

c. The genotypes of the two lines should be *A/A ; b/b* and *a/a ; B/B*.

d. The F_1 would be all be pigmented, *A/a ; B/b*. This is an example of complementation. The mutants are defective for different genes. The F_2 would be

9	*A/– ; B/–*	Pigmented
3	*a/a ; B/–*	White
3	*A/– ; b/b*	White
1	*a/a ; b/b*	White

9. **a.** Intercrossing mutant strains that all share a common recessive phenotype is the basis of the complementation test. This test is designed to identify the number of different genes that can mutate to a particular phenotype. If the progeny of a given cross still express the mutant phenotype, the mutations fail to complement and are considered alleles of the same gene; if the progeny are wild type, the mutations complement and the two strains carry mutant alleles of separate genes.

b. There are 3 genes represented in these crosses. All crosses except 2 × 3 (or 3 × 2) complement and indicate that the strains are mutant for separate genes. Strains 2 and 3 fail to complement and are mutant for the same gene.

c. Let *A* and *a* represent alleles of gene 1; *B* and *b* represent alleles of gene 4; and c^2, c^3, and *C* represent alleles of gene 3.

Line 1: *a/a . B/B . C/C*
Line 2: *A/A . B/B .* c^2/c^2
Line 3: *A/A . B/B .* c^3/c^3
Line 4: *A/A . b/b . C/C*

	Cross	Genotype	Phenotype
F_1s	1 × 2	*A/a . B/B .* C/c^2	Wild type
	1 × 3	*A/a . B/B .* C/c^3	Wild type
	1 × 4	*A/a . B/b . C/C*	Wild type
	2 × 3	*A/A . B/B .* c^2/c^3	Mutant
	2 × 4	*A/A . B/b .* C/c^2	Wild type
	3 × 4	*A/A . B/b .* C/c^3	Wild type

d. With the exception that strain 2 and 3 fail to complement and therefore have mutations in the same gene, this test does not give evidence of linkage. To test linkage, the F_1s should be crossed to tester strains (homozygous recessive strains) and segregation of the mutant phenotype followed. If the genes are unlinked, for example *A/a ; B/b* × *a/a ; b/b*, then 25% of the progeny will be wild type (*A/a ; B/b*) and 75% will be mutant (25% *A/a ; b/b*, 25% *a/a ; B/b*, and 25% *a/a ; b/b*). If the genes are linked (*a B/a B* × *A b/A b*), then only one half of the recombinants (i.e., less than 25% of the total progeny) will be wild type (*A B/a b*).

e. No. All it tells you is that among these strains, there are three genes represented. If genetic dissection of leg coordination was desired, large screens for the mutant phenotype would be executed with the attempt to "saturate" the genome with mutations in all genes involved in the process.

10. a. Complementation refers to gene products within a cell, which is not what is happening here. Most likely, what is known as cross-feeding is occurring, whereby a product made by one strain diffuses to another strain and allows growth of the second strain. This is equivalent to supplementing the medium. Because cross-feeding seems to be taking place, the suggestion is that the strains are blocked at different points in the metabolic pathway.

b. For cross-feeding to occur, the growing strain must have a block that occurs earlier in the metabolic pathway than does the block in the strain from which it is obtaining the product for growth.

c. The *trpE* strain grows when cross-fed by either *trpD* or *trpB* but the converse is not true (placing *trpE* earlier in the pathway than either *trpD* or *trpB*), and *trpD* grows when cross-fed by *trpB* (placing *trpD* prior to *trpB*). This suggests that the metabolic pathway is

$$trpE \rightarrow trpD \rightarrow trpB$$

d. Without some tryptophan, no growth at all would occur, and the cells would not have lived long enough to produce a product that could diffuse.

11. a. This is an example where one phenotype in the parents gives rise to three phenotypes in the offspring. The "frizzle" is the heterozygous phenotype and shows incomplete dominance.

P $\quad$ *A/a* (frizzle) × *A/a* (frizzle)

F_1 $\quad$ 1 *A/A* (normal):2 *A/a* (frizzle):1 *a/a* (woolly)

b. If *A/A* (normal) is crossed to *a/a* (woolly), all offspring will be *A/a* (frizzle).

12. a. The best explanation is that Marfan's syndrome is inherited as a dominant autosomal trait since roughly half of the children of all affected individuals also express the trait. If it were recessive, then all individuals marrying affected spouses would have to be heterozygous for an allele that when homozygous causes Marfan's.

b. The pedigree shows both pleiotropy (multiple affected traits) and variable expressivity (variable degree of expressed phenotype). Penetrance is the percentage of individuals with a specific genotype who express the associated phenotype. There is no evidence of decreased penetrance in this pedigree.

c. Pleiotropy indicates that the gene product is required in a number of different tissues, organs, or processes. When the gene is mutant, all tissues needing the gene product will be affected. Variable expressivity of a phenotype for a given genotype indicates modification by one or more other genes, random noise, or environmental effects.

13. It is possible to produce black offspring from two pure-breeding recessive albino parents if albinism results from mutations in either of two different genes. If the cross is designated

A/A . b/b × *a/a . B/B*

then all the offspring would be

A/a . B/b

and they would have a black phenotype because of complementation.

14. *Unpacking the Problem*

a. The character being studied is petal color.

b. The wild-type phenotype is blue.

c. A variant is a phenotypic difference from wild type that is observed.

d. There are two variants: pink and white.

e. "In nature" means that the variants did not appear in laboratory stock and, instead, were found growing wild.

f. Possibly the variants appeared as a small patch or even a single plant within a larger patch of wild type.

g. Seeds would be grown to check the outcome from each cross.

h. Given that no sex linkage appears to exist (sex is not specified in parents or offspring), "blue × white" means the same as "white × blue." Similar results would be expected because the trait being studied appears to be autosomal.

i. The first two crosses show a 3:1 ratio in the F_2, suggesting the segregation of one gene. The third cross has a 9:4:3 ratio for the F_2, suggesting that two genes are segregating.

j. Blue is dominant to both white and pink.

k. *Complementation* refers to generation of wild-type progeny from the cross of two strains that are mutant in different genes.

l. The ability to make blue pigment requires two enzymes that are individually defective in the pink or white strains. The F_1 progeny of this cross is blue, since each has inherited one nonmutant allele for both genes and can therefore produce both functional enzymes.

m. Blueness from a pink × white cross arises through complementation.

n. The following ratios are observed: 3:1, 9:4:3.

o. There are monohybrid ratios observed in the first two crosses.

p. There is a modified 9:3:3:1 ratio in the third cross.

q. A monohybrid ratio indicates that one gene is segregating, while a dihybrid ratio indicates that two genes are segregating.

r. 15:1, 12:3:1, 9:6:1, 9:4:3, 9:7

s. There is a modified dihybrid ratio in the third cross.

t. A modified dihybrid ratio most frequently indicates the interaction of two or more genes.

u. Recessive epistasis is indicated by the modified dihybrid ratio.

v.

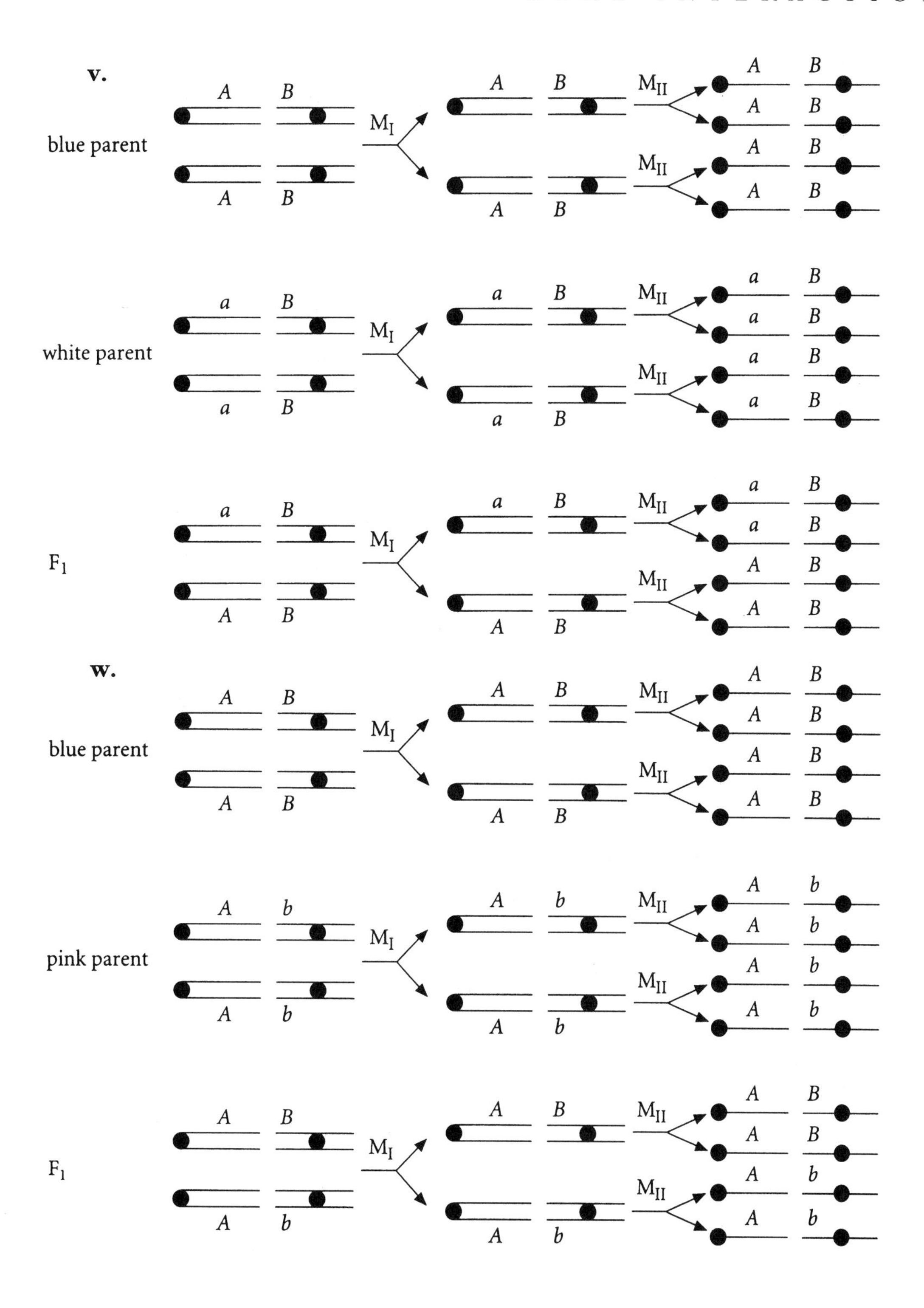
blue parent
white parent
F1
w.
blue parent
pink parent
F1
A
B
a
b
MI
MII

Solution to the Problem

a. Let A = wild type, a = white, B = wild type, and b pink.

Cross 1:	P	Blue × white	A/A ; $B/B \times a/a$; B/B
	F_1	All blue	All A/a ; B/B
	F_2	3 Blue:1 white	3 $A/–$; B/B:1 a/a ; B/B
Cross 2:	P	Blue × pink	A/A ; $B/B \times A/A$; b/b
	F_1	All blue	All A/A ; B/b
	F_2	3 Blue:1 pink	3 A/A ; $B/–$:1 A/A ; b/b
Cross 3:	P	pink × white	A/A ; $b/b \times a/a$; B/B
	F_1	All blue	All A/a ; B/b
	F_2	9 blue	9 $A/–$; $B/–$
		4 white	3 a/a ; $B/–$:1 a/a ; b/b
		3 pink	3 $A/–$; b/b

When the allele a is homozygous, the expression of alleles B or b is blocked or masked. The white phenotype is epistatic to the pigmented phenotypes. It is likely that the product of the A gene produces an intermediate that is then modified by the product of the B gene. If the plant is a/a, this intermediate is not made and the phenotype of the plant is the same regardless of the ability to produce functional B product.

b. The cross is

F_2	Blue × white
F_3	$^3/_8$ blue
	$^1/_8$ pink
	$^4/_8$ white

Begin by writing as much of each genotype as can be assumed.

F_2	$A/–$; $B/– \times a/a$; $–/–$
F_3	$^3/_8$ $A/–$; $B/–$
	$^1/_8$ $A/–$; b/b
	$^4/_8$ a/a ; $–/–$

Notice that both a/a and b/b appear in the F_3 progeny. In order for these homozygous recessives to occur, each parent must have at least one a and one b. Using this information, the cross becomes

F_2	A/a ; $B/b \times a/a$; $b/–$
F_3	$^3/_8$ A/a ; B/b
	$^1/_8$ A/a ; b/b
	$^4/_8$ a/a ; $b/–$

The only remaining question is whether the white parent was homozygous recessive, b/b, or heterozygous, B/b. If the white parent had been homozygous recessive, then the cross would have been a testcross of the blue parent, and the progeny ratio would have been 1 blue:1 pink:2 white, or 1 A/a ; B/b:1 A/a ; b/b:1 a/a ; B/b:1 a/a ; b/b. This was not observed. Therefore, the white parent had to have been heterozygous, and the F_2 cross was A/a ; $B/b \times a/a$; B/b.

15. a., b. Crosses 1–3 show a 3:1 ratio, indicating that brown, black, and yellow are all alleles of one gene. Crosses 4–6 show a modified 9:3:3:1 ratio, indicating that at least two genes are involved. Those crosses also indicate that the presence of color is dominant to its absence. Furthermore, epistasis must be involved for there to be a modified 9:3:3:1 ratio.

By looking at the F_1 of crosses 1–3, the following allelic dominance relationships can be seen easily: black > brown > yellow. Arbitrarily assign the following genotypes for homozygotes: B^l/B^l = black, B^r/B^r = brown, B^y/B^y = yellow.

By looking at the F_2 of crosses 4–6, a white phenotype is composed of two categories: the double homozygote and one class of the mixed homozygote/heterozygote. Let lack of color be caused by *c/c*. Color will therefore be *C/–*.

Parents	F_1	F_2
1 B^r/B^r ; C/C × B^y/B^y ; C/C	B^r/B^y ; C/C	3 $B^r/–$; C/C:1 B^y/B^y ; C/C
2 B^l/B^l ; C/C × B^r/B^r ; C/C	B^l/B^r ; C/C	3 $B^l/–$; C/C:1 B^r/B^r ; C/C
3 B^l/B^l ; C/C × B^y/B^y ; C/C	B^l/B^y ; C/C	3 $B^l/–$; C/C:1 B^y/B^y ; C/C
4 B^l/B^l ; c/c × B^y/B^y ; C/C	B^l/B^y ; C/c	9 $B^l/–$; $C/–$:3 B^y/B^y; $C/–$: 3 $B^l–$; c/c: 1 B^y/B^y; c/c
5 B^l/B^l ; c/c × B^r/B^r ; C/C	B^l/B^r ; C/c	9 $B^l/–$; $C/–$:3 B^r/B^r ; $C/–$:3 $B^l/–$; c/c: 1 B^r/B^r ; c/c
6 B^l/B^l ; C/C × B^y/B^y ; c/c	B^l/B^y ; C/c	9 $B^l/–$; $C/–$:3 B^y/B^y; $C/–$:3 $B^l/–$; c/c: 1 B^y/B^y ; c/c

c. The following biochemical pathway is suggested by the data

White —(*c*)→ yellow —(B^y)→ brown —(B^r)→ black

16. To solve this problem, first restate the information.

A/–	yellow	*A/–* ; *R/–*	gray
R/–	black	*a/a* ; *r/r*	white

The cross is gray × yellow, or *A/–* ; *R/–* × *A/–* ; *r/r*. The F_1 progeny are

$^3/_8$ yellow	$^1/_8$ black
$^3/_8$ gray	$^1/_8$ white

For white progeny, both parents must carry an *r* and an *a* allele. Now the cross can be rewritten as: *A/a* ; *R/r* × *A/a* ; *r/r*

17. a. The stated cross is

P Single-combed (*r/r* ; *p/p*) × walnut-combed (*R/R* ; *P/P*)

F_1	*R/r* ; *P/p*	Walnut
F_2	9 *R/–* ; *P/–*	Walnut
	3 *r/r* ; *P/–*	Pea
	3 *R/–* ; *p/p*	Rose
	1 *r/r* ; *p/p*	Single

b. The stated cross is

P Walnut-combed × rose-combed

and the F_1 progeny are

	Phenotypes	Possible genotypes
3/8	rose	*R/– ; p/p*
3/8	walnut	*R/– ; P/–*
1/8	pea	*r/r ; P/–*
1/8	single	*r/r ; p/p*

The 3 *R/–*:1 *r/r* ratio indicates that the parents were heterozygous for the *R* gene. The 1 *P/–*:1 *p/p* ratio indicates a testcross for this gene. Therefore, the parents were *R/r ; P/p* and *R/r ; p/p*.

c. The stated cross is

P Walnut-combed × rose-combed

F_1 Walnut (*R/– ; P/p*)

To get this result, one of the parents must be homozygous *R*, but both need not be, and the walnut parent must be homozygous *P/P*.

d. The following genotypes produce the walnut phenotype

R/R ; P/P, *R/r ; P/P*, *R/R ; P/p*, *R/r ; P/p*

18. a. This type of gene interaction is called *epistasis*. The phenotype of *e/e* is epistatic to the phenotypes of *B/–* or *b/b*.

b. The progeny of generation I have all possible phenotypes. Progeny II-3 is beige (*e/e*), so both parents must be heterozygous *E/e*. Progeny II-4 is brown (*b/b*), so both parents must also be heterozygous *B/b*. Progeny III-3 and III-5 are brown, so II-2 and II-5 must be *B/b*. Progeny III-2 and III-7 are beige (*e/e*), so all their parents must be *E/e*.

The following are the inferred genotypes:

I 1 (*B/b E/e*) 2 (*B/b E/e*)
II 1 (*b/b E/e*) 2 (*B/b E/e*) 3 (*–/– e/e*) 4 (*b/b E/–*) 5 (*B/b E/e*) 6 (*b/b E/e*)
III 1 (*B/b E/–*) 2 (*–/b e/e*) 3 (*b/b, E/–*) 4 (*B/b E/–*) 5 (*b/b E/–*) 6 (*B/b E/–*) 7 (*–/b e/e*)

19. a. Note that blue is always present, indicating *E/E* (blue) in both parents. Because of the ratios that are observed, neither *C* nor *D* is varying. In this case, the gene pairs that are involved are *A/a* and *B/b*. The parents are *A/A ; b/b* × *a/a ; B/B* or *A/A ; B/B* × *a/a ; b/b*.

The F_1 are *A/a ; B/b* and the F_2 are

9	*A/– ; B/–*	Blue + red, or purple
3	*A/– ; b/b*	Blue + yellow, or green
3	*a/a ; B/–*	Blue + $white_2$, or blue
1	*a/a ; b/b*	Blue + $white_2$, or blue

b. Blue is not always present, indicating *E/e* in the F_1. Because green never appears, the F_1 must be *B/B . C/C . D/D*. The parents are *A/A ; e/e* × *a/a ; E/E* or *A/A ; E/E* × *a/a ; e/e*.

The F_1 are *A/a ; E/e*, and the F_2 are

9	*A/– ; E/–*	Red + blue, or purple
3	*A/– ; e/e*	Red + $white_1$, or red
3	*a/a ; E/–*	$White_2$ + blue, or blue
1	*a/a ; e/e*	$White_2$ + $white_1$, or white

c. Blue is always present, indicating that the F_1 is *E/E*. No green appears, indicating that the F_1 is also *B/B*. The two genes involved are *A* and *D*. The parents are *A/A ; d/d* × *a/a ; D/D* or *A/A ; D/D* × *a/a ; d/d*.

The F_1 are *A/a ; D/d* and the F_2 are:

9	*A/– ; D/–*	Blue + red + $white_4$, or purple
3	*A/– ; d/d*	Blue + red, or purple
3	*a/a ; D/–*	Blue + $white_2$ + $white_4$, or blue
1	*a/a ; d/d*	$White_2$ + blue + red, or purple

d. The presence of yellow indicates *b/b ; e/e* in the F_2. Therefore, the parents are *B/B ; e/e* × *b/b ; E/E* or *B/B ; E/E* × *b/b ; e/e*.

The F_1 are *B/b ; E/e* and the F_2 are:

9	*B/– ; E/–*	Red + blue, or purple
3	*B/– ; e/e*	Red + $white_1$, or red
3	*b/b ; E/–*	Yellow + blue, or green
1	*b/b ; e/e*	Yellow + $white_1$, or yellow

e. Mutations in *D* suppress mutations in *A*.

f. Recessive alleles of *A* will be epistatic to mutations in *B*.

20. a. The trait is recessive (parents without the trait have children with the trait) and autosomal (daughters can inherit the trait from unaffected fathers). Looking to generation III, there is also evidence that there are two different genes that when defective result in deaf-mutism.

Assuming that one gene has alleles *A* and *a* and the other has *B* and *b*, the following genotypes can be inferred:

I-1 and I-2	*A/a ; B/B*	I-3 and I-4	*A/A ; B/b*
II-(1, 3, 4, 5, 6)	*A/– ; B/B*	II- (9, 10, 12, 13, 14, 15)	*A/A ; B/–*
II-2 and II-7	*a/a ; B/B*	II-8 and II-11	*A/A ; b/b*

b. Generation III shows complementation. All are *A/a ; B/b*.

21. a. The first impression from the pedigree is that the gene causing blue sclera and brittle bones is pleiotropic with variable expressivity. If two genes were involved, it would be highly unlikely that all people with brittle bones also had blue sclera.

b. Sons and daughters inherit from affected fathers so the allele appears to be autosomal.

c. The trait appears to be inherited as a dominant but with incomplete penetrance. For the trait to be recessive, many of the nonrelated individuals marrying into the pedigree would have to be heterozygous (e.g., I-1, I-3,

II-8, II-11). Individuals II-4, II-14, III-2, and III-14 have descendants with the disorder although they do not themselves express the disorder. Therefore, $^4/_{20}$ people that can be inferred to carry the gene, do not express the trait. That is 80% penetrance. (Penetrance could be significantly less than that since many possible carriers have no shown progeny.) The pedigree also exhibits variable expressivity. Of the 16 individuals who have blue sclera, 10 do not have brittle bones. Usually, expressivity is put in terms of none, variable, and highly variable, rather than expressed as percentages.

22. The first two crosses indicate that wild type is dominant to both platinum and aleutian. The third cross indicates that two genes are involved rather than one gene with multiple alleles because a 9:3:3:1 ratio is observed.

Let platinum be *a*, aleutian be *b*, and wild type be *A/–* ; *B/–*.

Cross 1:	P	*A/A* ; *B/B* × *a/a* ; *B/B*	Wild type × platinum
	F_1	*A/a* ; *B/B*	All wild type
	F_2	3 *A/–* ; *B/B*:1 *a/a* ; *B/B*	3 wild type:1 platinum
Cross 2:	P	*A/A* ; *B/B* × *A/A* ; *b/b*	Wild type × aleutian
	F_1	*A/A* ; *B/b*	All wild type
	F_2	3 *A/A* ; *B/–*:1 *A/A* ; *b/b*	3 wild type:1 aleutian
Cross 3:	P	*a/a* ; *B/B* × *A/A* ; *b/b*	Platinum × aleutian
	F_1	*A/a* ; *B/b*	All wild type
	F_2	9 *A/–* ; *B/–*	Wild type
		3 *A/–* ; *b/b*	Aleutian
		3 *a/a* ; *B/–*	Platinum
		1 *a/a* ; *b/b*	Sapphire

b.

	Sapphire × platinum		Sapphire × aleutian	
P	*a/a* ; *b/b* × *a/a* ; *B/B*		*a/a* ; *b/b* × *A/A* ; *b/b*	
F_1	*a/a* ; *B/b*	Platinum	*A/a* ; *b/b*	Aleutian
F_2	3 *a/a* ; *B/–*	Platinum	3 *A/–* ; *b/b*	Aleutian
	1 *a/a* ; *b/b*	Sapphire	1 *a/a* ; *b/b*	Sapphire

23. a. The first experiment is a complementation test. This test is designed to identify the number of different genes involved in the analysis. In this problem, if the heterokaryon still cannot grew in the absence of leucine, the mutations fail to complement and are considered alleles of the same gene; if the heterokaryons grow, the mutations complement and the two strains carry mutant alleles of separate genes.

The second experiment is a recombination test to indicate whether the genes are linked. If the genes are unlinked, for example *A/a* ; *B/b* × *a/a* ; *b/b*, then 25% of the progeny will be wild type (*A/a* ; *B/b*) and 75% will be mutant (25% *A/a* ; *b/b*, 25% *a/a* ; *B/b*, and 25% *a/a* ; *b/b*). If the genes are linked (*a B/a B* × *A b/A b*), then only one half of the recombinants (i.e., less than 25% of the total progeny) will be wild type (*A B/a b*).

b. For all heterokaryons except those from *a* × *e*, growth in the absence of leucine occurs. This indicates that *a* and *e* are mutations in the same gene and that b, c, and d are mutations in separate genes. In other words, four genes are being analyzed.

c. In the second experiment, if the genes are unlinked, 25% of the progeny should be leucine-independent. The frequency of prototrophs is approximately 25% in all pairwise crosses except for $a \times e$ (0 prototrophic progeny) and $b \times d$ (2 prototrophic progeny). From the complementation test, it is already known that a and e are in the same gene. This test now indicates that b and d are linked and that the RF = 100% × 4/500 = 0.8 map units. (Remember, the prototrophs represent only one half of the recombinants. The other recombinants are double mutants that will not be detected.)

d.

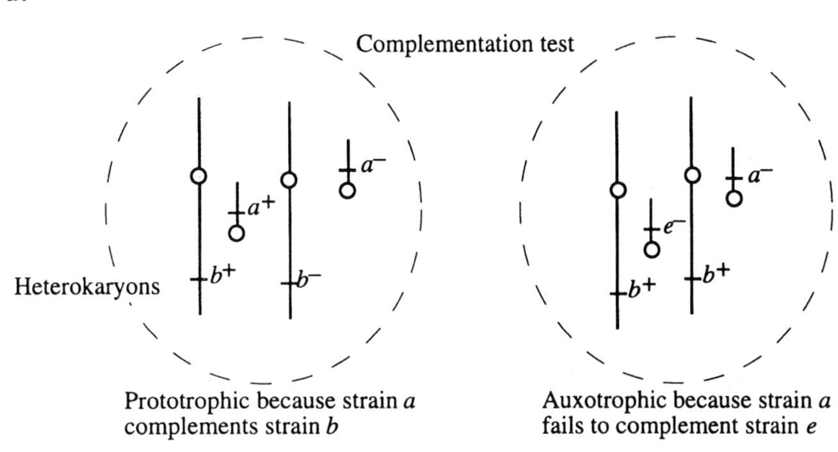

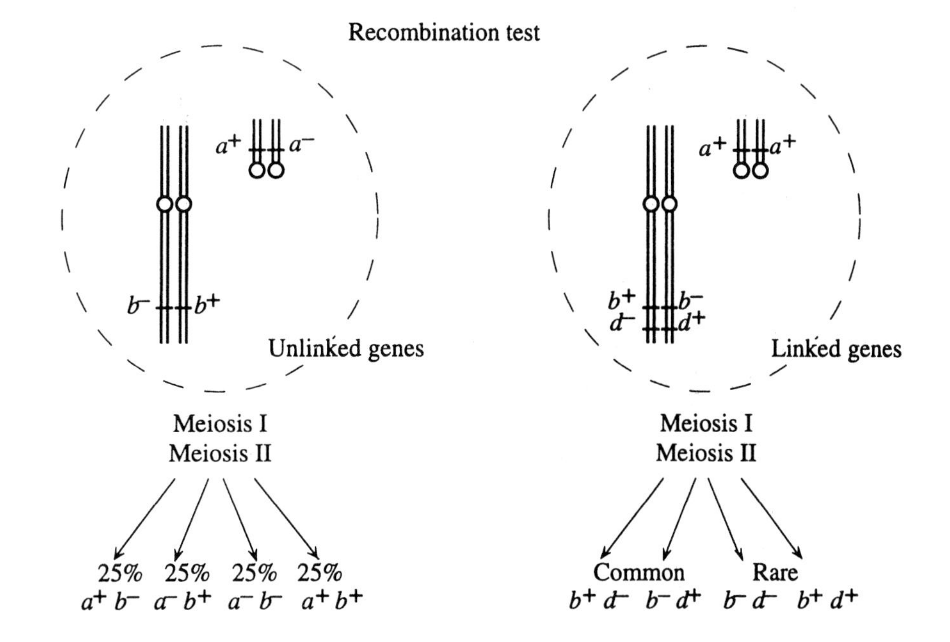

24. a. There are 10 mutants that fall into three complementation groups. Mutants 1, 2, 7 fail to complement; 3, 5, 8 fail to complement; and 4, 6, 9, 10 fail to complement. There are three genes indicated by this complementation test.

b. There is evidence of a biochemical pathway. If a mutant grows on a supplement, the defect is earlier in the pathway; if a mutant fails to grow when supplemented, the defect is later in the pathway. Mutant 1 grows on SAICAR but not CAIR or AIR. The step blocked in mutant 1 is before SAICAR but after CAIR and AIR. Similarly, the step blocked in mutant 3 is after SAICAR, CAIR, and AIR and the step blocked in mutant 4 is before CAIR and SAICAR but after AIR. The indicated pathway is

$$\text{AIR} \xrightarrow{4} \text{CAIR} \xrightarrow{1} \text{SAICAR} \xrightarrow{3} \text{Adenine}$$

Where a mutant is blocked is indicated by the vertical line through the arrow.

c. Ten mutations that by complementation represent three genes involved in the metabolism of adenine are being studied. The recombination test indicates that the genes represented by mutants 1 and 3 are linked. The gene represented by mutant 4 is not linked to the other two. For unlinked genes, approximately 25% of the progeny should be prototrophic, and this is true for crosses 1 × 4 and 3 × 4. Cross 1 × 3 (and 3 × 1) showed 11 prototrophs out of a possible 2000 progeny. This represents half of the possible recombinants (the other half being doubly mutant) and indicates an RF = 100% × 22/2000 = 1.1%.

25. The cross is e^+/e ; r^+/r × e^+/e ; r^+/r. For wild-type function, a functional regulatory protein and a functional gene encoding the enzyme is required. The progeny are

$^9/_{16}$	$e^+/-$; $r^+/-$	Have enzyme
$^3/_{16}$	$e^+/-$; r/r	Lack enzyme
$^3/_{16}$	e/e ; $r^+/-$	Lack enzyme
$^1/_{16}$	e/e ; r/r	Lack enzyme

26. a. The data from the crosses indicate that the mutations are in different, unlinked genes and both are recessive. The data from the gel indicate that one mutation is in the gene that codes for the P protein. As a result, the mutant protein is truncated (smaller) and runs more rapidly in the gel. Possible mutations giving this result might include nonsense, frameshifts, deletions, or mutations leading to altered splicing. (Alternatively, you could conclude that the mutation causes a larger, but still nonfunctional

P protein to be made due to an insertion or splicing defect.) The other mutation is in a gene that codes for a regulatory protein required for P gene expression. When mutant, P expression is greatly reduced.

b. Assume p^+ = normal P protein; r^+ = normal regulatory protein; both are required for normal function.

Lane 1:	p^+/p^+ ; $r^+/-$
Lane 2:	p^+/p^+ ; r/r
Lane 3:	p^+/p ; $r^+/-$
Lane 4:	p^+/p ; r/r
Lane 5:	p/p ; $r^+/-$
Lane 6:	p/p ; r/r

c. Type 4 is p^+/p ; r/r, $^1/_2$ of the progeny will be p^+/p, and $^1/_4$ of the progeny will be independently r/r, so $^1/_2 \times {^1/_4} = {^1/_8}$.

d. Lane 1 (p^+/p^+ ; $r^+/-$) and lane 3 (p^+/p ; $r^+/-$) will be phenotypically wild type.

e. Parents, p/p ; r^+/r^+ × p^+/p^+ ; r/r will look like lane 5 and lane 2, respectively. F_1, p^+/p ; r^+/r will look like lane 3.

27. The three mutations are recessive. Mutant line 1 and line 2 are mutations in the same gene (they do not complement), and mutant line 1 and line 3 are mutations in different genes (they do complement). Assume line 1 is m^1/m^1; line 2 is m^2/m^2; and line 3 is m^3/m^3.

Line 1: $m^1/m^1 \times +/+$ → all $m^1/+$ → self → $^3/_4$ $+/-$:$^1/_4$ m^1/m^1

Line 2: $m^2/m^2 \times +/+$ → all $m^2/+$ → self → $^3/_4$ $+/-$:$^1/_4$ m^2/m^2

Line 3: $m^3/m^3 \times +/+$ → all $m^3/+$ → self → $^3/_4$ $+/-$:$^1/_4$ m^3/m^3

Line 1 × Line 2

$m^1/m^1 \times m^2/m^2$

↓

All: m^1/m^2

↓

Self: 1 m^1/m^1
2 m^1/m^2
1 m^2/m^2

Line 1 × Line 3

m^1/m^1 ; $+/+$ × $+/+$; m^3/m^3

↓

All: $m^1/+$; $m^3/+$

↓

Self:
$^9/_{16}$ $+/-$; $+/-$
$^3/_{16}$ $+/-$; m^3/m^3
$^3/_{16}$ m^1/m^1 ; $+/-$
$^1/_{16}$ m^1/m^1 ; m^3/m^3

28. These data suggest that three alleles of one gene are being studied and that the order of dominance is black (*B*) > red (*r*) > blue (*bl*).

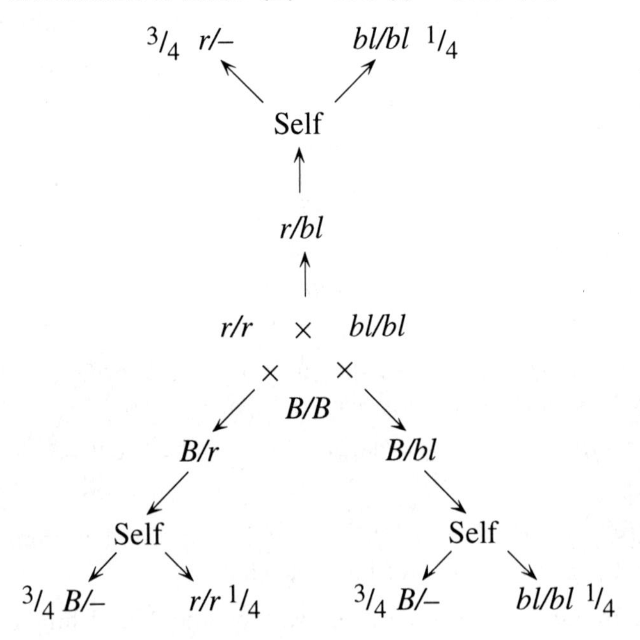

29. A single gene showing incomplete dominance (1:2:1 ratio in F_2)

A/A × *a/a*
↓
25% *A/A*
50% *A/a*
25% *a/a*

30. There is a 13:3 ratio of F_2 progeny suggesting two genes segregating independently and epistasis.

Pure-breeding × Pure-breeding
A/A ; *b/b* × *a/a* ; *B/B*
↓
All *A/a* ; *B/b*
↓
Self
↓

131 9 *A/–* ; *B/–*:3 *a/a* ; *B/–*:1 *a/a* ; *b/b*
29 3 *A/–* ; *b/b*

31. There is a 12:3:1 ratio of F_2 progeny suggesting two genes segregating independently and epistasis.

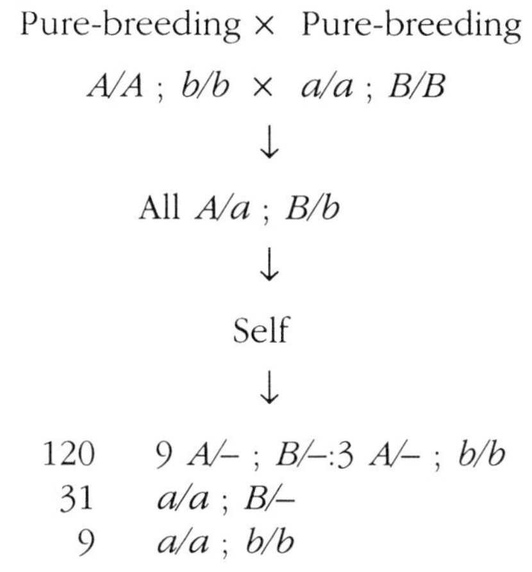

32. These data suggest that three alleles of one gene are being studied. Both codominance (=) and classical dominance (>) are present in this multiple allelic series: l^v (vertical lines) = l^h (horizontal lines), and both l^v and $l^h > l^o$ (no lines).

$l^v/l^v \times l^o/l^o \rightarrow$ all $l^v/l^o \rightarrow F_2$ 3 $l^v/–$:1 l^o/l^o
$l^h/l^h \times l^o/l^o \rightarrow$ all $l^h/l^o \rightarrow F_2$ 3 $l^h/–$:1 l^o/l^o
$l^v/l^v \times l^h/l^h \rightarrow$ all $l^v/l^h \rightarrow F_2$ 1 l^v/l^v:2 l^v/l^h:1 l^h/l^h

33. **a.** The results indicate that two unlinked genes are being studied. One gene determines whether the beetles have a diamond or stripe. For these phenotypes, the diamond allele (*D*) is dominant to the stripe allele (*d*). The other gene determines whether the beetle has a spot. In this case, the spot allele (*S*) is dominant to the no-spot allele (*s*), and it is also epistatic to *D* and *d*.

Cross 1
D/d ; *s/s* × *D/d* ; *s/s* → $^1/_4$ *D/D* ; *s/s* + $^1/_2$ *D/d* ; *s/s* (diamond)
$^1/_4$ *d/d* ; *s/s* (stripe)

Cross 2
D/d ; *S/s* × *d/d* ; *S/s* → $^1/_8$ *D/d* ; *S/S* + $^1/_8$ *d/d* ; *S/S* + $^1/_4$ *D/d* ; *S/s* + $^1/_4$ *d/d* ; *S/s* (spot)
$^1/_8$ *D/d* ; *s/s* (diamond)
$^1/_8$ *d/d* ; *s/s* (stripe)

Cross 3
D/d ; *S/s* × *d/d* ; *s/s* → $^1/_4$ *d/d* ; *S/s* + $^1/_4$ *D/d* ; *S/s* (spot)
$^1/_4$ *D/d* ; *s/s* (diamond)
$^1/_4$ *d/d* ; *s/s* (stripe)

b. The spotted progeny represent two genotypes

$^1/_2$ *D/d* ; *S/s*
$^1/_2$ *d/d* ; *S/s*

When intercrossed, there are three possible combinations:

(a) $^1/_4$ *D/d* ; *S/s* × *D/d* ; *S/s*
(b) $^1/_2$ *D/d* ; *S/s* × *d/d* ; *S/s*
(c) $^1/_4$ *d/d* ; *S/s* × *d/d* ; *S/s*

From (a) $^3/_4$ spot; $^1/_{16}$ stripe; $^3/_{16}$ diamond (× $^1/_4$ of total crosses)
(b) $^3/_4$ spot; $^1/_8$ stripe; $^1/_8$ diamond (× $^1/_2$ of total crosses)
(c) $^3/_4$ spot; $^1/_4$ stripe (× $^1/_4$ of total crosses)

Weighted total: $^3/_4$ spot; $^9/_{64}$ stripe; $^7/_{64}$ diamond

7 Gene Mutations

1. All cells derived from the cell in which the reversion took place will now be w^+/w. Depending on when this took place, the petal will now be blue, either in part or in whole. Since the petal is part of the plant's soma, this reversion would not be inherited.

2. Starting with a yeast strain that is *pro-1*, plate the cells on medium lacking proline. Only those cells that are able to synthesize proline will form colonies. Most of these will be revertants; however, some will have second-site suppressors. (Treating the cells with a mutagen prior to plating them would significantly increase the yield.)

3. There are many ways to test a chemical for mutagenicity. For example, Figure 7-36 of the text discusses a detection system for recessive somatic mutations in mice. Mice bred to be heterozygous for seven genes involved in coat color are exposed to a potential mutagen by injecting it into the uterus of their pregnant mother. Any somatic mutation from wild type to mutant at one of the seven loci will result in a patch (or mutant sector) of differently colored fur. The number of mutant sectors later found on these chemically treated mice would be compared with the number found on genetically identical, but chemically untreated, mice (the control mice). (A proper control would expose the control mice to exactly the same experimental protocol except for the caffeine. This would include injection of whatever solvent was used into the uterus of their mother at the same developmental time as for the experimental mice.)

Much larger screens or selections could be done with fungi. For example, haploid *Neurospora* auxotrophic for the amino acid leucine could be exposed to caffeine and then plated onto minimal medium selecting for leu^+ colonies. Only those cells in which a reverse mutation (from leu^- to leu^+) occurred would grow. Reversion rates of treated and untreated (control) cells would be compared to see if the caffeine was mutagenic. Alternatively, yeast cells could be exposed to caffeine and mutations in the gene *ade-3* could be scored and numbers compared between treated and untreated populations. (Mutations in this gene actually cause the yeast to be red instead of white, so large numbers of colonies can be screened rapidly.)

Although this question asks specifically for mutations in higher organisms, a rapid and widely used mutagen-detection system using bacteria was developed by Bruce Ames in the 1970s. Using a genetically modified bacterium *(Salmonella typhimurium)* that is auxotrophic for histidine and defective in DNA repair, the Ames test quickly ascertains the mutagenicity of various chemicals. Since the basic properties of DNA and mutation are the same in prokaryotes and eukaryotes, this test does have relevance. The Ames test has been further enhanced by using rat liver extracts to modify the tested chemicals to simulate human (and mammalian) metabolism. This is important because although the liver is responsible for most of the detoxification and metabolism of ingested chemicals (hence the connection between alcohol and liver disease!), some chemicals are modified in ways that actually make them toxic or mutagenic.

4. The mutation rate needs to be corrected for achondroplastic parents and put on a "per gamete" basis. Mutation rate = $(10 - 2)/[2 \times (94{,}075 - 2)] = 4.25 \times 10^{-5}$.

For this problem, you do not have to worry about revertants since you are asked only for the net mutation frequency.

5. The commission was looking for induced recessive X-linked lethal mutations, which would show up as a shift in the sex ratio. A shift in the sex ratio is the first indication that a population has sustained lethal genetic damage. Other recessive mutations might have occurred, of course, but they would not be homozygous and therefore would go undetected. All dominant mutations would be immediately visible, unless they were lethal. If they were lethal, there would be lowered fertility, an increase in detected abortions, or both, but the sex ratio would not shift as dramatically.

6. The mutants can be categorized as follows:
Mutant 1: an auxotrophic mutant
Mutant 2: a non-nutritional, temperature-sensitive mutant
Mutant 3: a leaky, auxotrophic mutant
Mutant 4: a leaky, non-nutritional, temperature-sensitive mutant
Mutant 5: a non-nutritional, temperature-sensitive, auxotrophic mutant

7. **a.** A transition mutation is the substitution of a purine for a purine or the substitution of a pyrimidine for a pyrimidine. A transversion mutation is the substitution of a purine for a pyrimidine, or vice versa.

b. Both are base-pair substitutions. A silent mutation is one that does not alter the function of the protein product from the gene, because the new codon codes for the same amino acid as did the nonmutant codon. A neutral mutation results in a different amino acid that is functionally equivalent, and the mutation therefore has no adaptive significance.

c. A missense mutation results in a different amino acid in the protein product of the gene. A nonsense mutation causes premature termination of translation, resulting in a shortened protein.

d. Frameshift mutations arise from addition or deletion of one or more bases in other than multiples of three, thus altering the reading frame for translation. Therefore, the amino acid sequence from the site of the mutation to the end of the protein product of the gene will be altered. Frameshift mutations can and often do result in premature stop codons in the new reading frame, leading to shortened protein products.

8. Frameshift mutations arise from addition or deletion of one or more bases in other than multiples of three. When translated, this will alter the reading frame and therefore the amino acid sequence from the site of the mutation to the end of the protein product. Also, frameshift mutations often result in premature stop codons in the new reading frame, leading to shortened protein products. A missense mutation changes only a single amino acid in the protein product.

9. The Streisinger model proposed that frameshifts arise when loops in single stranded regions are stabilized by slipped mispairing of repeated sequences. In the *lac* gene of *E. coli*, a four-base-pair sequence is repeated three times in tandem, and this is the site of a hot spot.

The sequence is 5′-CTGG CTGG CTGG-3′. During replication the DNA must become single stranded in short stretches for replication to occur. As the new strand is synthesized, transient disruptions of the hydrogen bonds holding the new and old strands together may be stabilized by the incorrect base pairing of bases that are now out of register by the length of the repeat, or in this case, a total of four bases. Depending on which strand, new or template, loops out with respect to the other, there will be an addition or deletion of four bases, as diagrammed below:

```
             T G
            C   G
5′-C T G G     C T G G-3′  ⟶        DNA synthesis
3′-G A C C —  G A C C  G A C C -5′
```

In this diagram, the upper strand looped out as replication was occurring. The loop is stabilized by base pairing. As replication continues at the 3′ end, an additional copy of CTGG will be synthesized, leading to an addition of four bases. This will result in a frameshift mutation.

10. In Problem 9, had the lower strand (the template) looped, the result would have been a deletion in the newly synthesized upper strand.

Misalignment of homologous chromosomes during recombination results in a duplication in one strand and a corresponding deletion in the other.

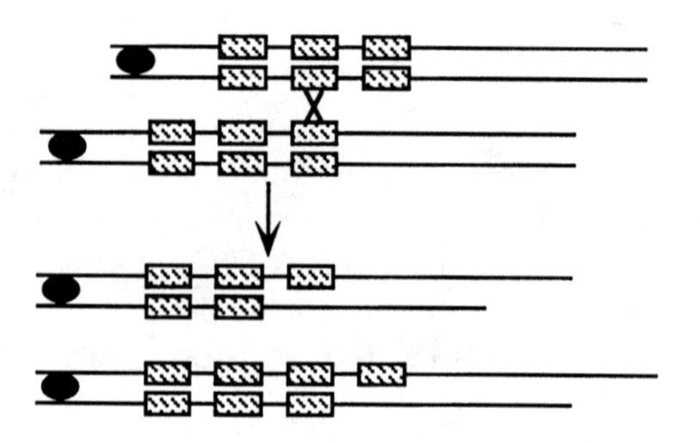

Recombination between two homologous repeats in a looped DNA molecule can lead to deletion.

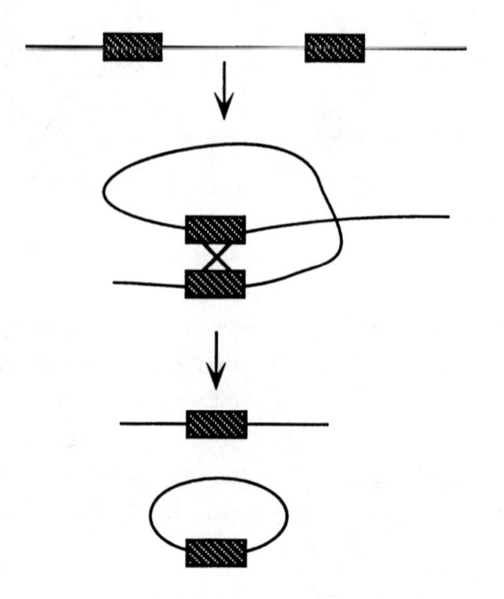

All these mechanisms are supported by DNA sequencing results.

11. Depurination results in the loss of adenine or guanine from the DNA. Since the resulting apurinic site cannot specify a complementary base, replication is blocked. Under certain conditions, replication proceeds with a near random insertion of a base opposite the apurinic site. In three-fourths of these insertions, a mutation will result.

Deamination of cytosine yields uracil. If left unrepaired, the uracil will be paired with adenine during replication, ultimately resulting in a transition mutation.

8-OxodG (8-oxo-7-hydrodeoxyguanosine) can pair with adenine, resulting in a transversion.

12. 5-BU is an analog of thymine. It undergoes tautomeric shifts at a higher frequency than does thymine and therefore is more likely to pair with G than thymine is during replication. At the next replication this will lead to a GC pair rather than the original AT pair. On the other hand, 5-BU can also be incorporated into DNA by mispairing with guanine. In this case it will convert a GC pair to an AT pair.

EMS is an alkylating agent that produces O-6-ethylguanine. This alkylated guanine will mispair with thymine, which leads from a GC pair to an AT pair at the next replication.

13. An AP site is an apurinic or apyrimidinic site. AP endonucleases introduce chain breaks by cleaving the phosphodiester bonds at the AP sites. Some exonuclease activity follows, so that a number of bases are removed. The resulting gap is filled by DNA pol I and then sealed by DNA ligase.

General excision repair is used to remove damaged DNA including photodimers. It cleaves the phosphodiester backbone on either side of the damage, removing 12 or 13 nucleotides in bacteria and 27 to 29 nucleotides in eukaryotes. The resulting gap is filled by DNA pol I and then sealed by DNA ligase.

Photodimers can also be repaired by photolyase (found in bacteria and lower eukaryotes), which binds to and splits the dimer in the presence of certain wavelengths of light. In bacteria the damaged DNA can also be bypassed by the SOS system, which enables DNA polymerase to "fill in" random bases where it encounters photodimers on the template strand. Finally, although photodimers on the template strand cause DNA pol III to stall, replication can restart downstream from the dimer, leaving a region of single-stranded DNA. The single-stranded DNA will attract single-stranded-binding protein and another protein, RecA, which can effect recombinational repair using DNA from the sister chromatid to patch the gap.

14. Mismatch repair occurs if a mismatched nucleotide is inserted during replication. The new, incorrect base is removed and the proper base is inserted. The enzymes involved can distinguish between new and old strands because, in *E. coli*, the old strand is methylated.

Recombination repair occurs if lesions such as AP sites and UV photodimers block replication (there is a gap in the new complementary strand). Recombination fills this gap with the corresponding segment from the sister DNA molecule, which is normal in both strands. This produces one DNA molecule with a gap across from a correct strand, which can then be filled by complementarity, and one with a photodimer across from a correct strand.

15. Leaky mutants are mutants with an altered protein product that retains a low level of function. Enzyme activity may, for instance, be reduced rather than abolished by a mutation.

16. a. You are told that all mutants are simple Mendelian recessives, so you need not worry about mutations that map to the chloroplast's genome. To determine the number of genes involved, a complementation test (conducting pairwise crosses) is performed. All combinations complement (are mutant for different genes) except for 1 × 4. In this cross, the F_1 is still mutant, the mutations fail to complement and are in the same gene. Thus, 3 genes are represented.

b. As stated above, 1 and 4 map to the same gene. Of the other combinations, only 2 and 3 show linkage. In this case, the testcross of the 2 × 3 F_1 produces 10% wild-type progeny, not the expected 25% if the genes were unlinked. You can also use these data to determine the map distance between genes 2 and 3. The percentage of wild-type progeny from the testcross will be equal to half that of the recombinants (the other half will be mutant for both genes). Thus genes 2 and 3 are 20 map units apart.

c. 1 × 2: m_1/m_1 ; +/+ × +/+ ; m_2/m_2

↓

$m_1/+$; $m_2/+$ × m_1/m_1 ; m_2/m_2 (testcross)

↓

25% m_1/m_1 ; $m_2/+$
25% $m_1/+$; m_2/m_2
25% m_1/m_1 ; m_2/m_2
25% $m_1/+$; $m_2/+$ (wild type)

1 × 3: m_1/m_1 ; +/+ × +/+ ; m_3/m_3

↓

$m_1/+$; $m_3/+$ × m_1/m_1 ; m_3/m_3 (testcross)

↓

25% m_1/m_1 ; $m_3/+$
25% $m_1/+$; m_3/m_3
25% m_1/m_1 ; m_3/m_3
25% $m_1/+$; $m_3/+$ (wild type)

1 × 4: m_1/m_1 × m_4/m_4

↓

m_1/m_4 × m_1/m_1 or m_4/m_4 (testcross)

↓

50% m_1/m_4
50% m_1/m_1

2 × 3: $m_2 +/m_2 +$ × $+ m_3/+ m_3$
↓
$m_2 +/+ m_3$ × $m_2 m_3/m_2 m_3$ (testcross)
↓
40% $m_2 +/m_2 m_3$
40% $+ m_3/m_2 m_3$
10% $m_2 m_3/m_2 m_3$
10% $+ +/m_2 m_3$ (wild type)

2 × 4: as in 1 × 2
3 × 4: as in 1 × 2

17. a. Because 5′-UAA-3′ does not contain G or C, a transition to a GC pair in the DNA cannot result in 5′-UAA-3′. 5′-UGA-3′ and 5′-UAG-3′ have the DNA antisense-strand sequence of 3′-ACT-5′ and 3′-ATC-5′, respectively. A transition to either of these stop codons occurs from the nonmutant 3′-ATT-5′, respectively. However, a DNA sequence of 3′-ATT-5′ results in an RNA sequence of 5′-UAA-3′, itself a stop codon.

b. Yes. An example is 5′-UGG-3′, which codes for Trp, to 5′-UAG-3′.

c. No. In the three stop codons the only base that can be acted upon is G (in UAG, for instance). Replacing the G with an A would result in 5′-UAA-3′, a stop codon.

18. a., b. Mutant 1: most likely a deletion. It could be caused by radiation.

Mutant 2: because proflavin causes either additions or deletions of bases and because spontaneous mutation can result in additions or deletions, the most probable cause was a frameshift mutation by an intercalating agent.

Mutant 3: 5-BU causes transitions, which means that the original mutation was most likely a transition. Because HA causes GC-to-AT transitions and HA cannot revert it, the original must have been a GC-to-AT transition. It could have been caused by base analogs.

Mutant 4: the chemical agents cause transitions or frameshift mutations. Because there is spontaneous reversion only, the original mutation must have been a transversion. X-irradiation or oxidizing agents could have caused the original mutation.

Mutant 5: HA causes transitions from GC-to-AT, as does 5-BU. The original mutation was most likely an AT-to-GC transition, which could be caused by base analogs.

c. The suggestion is a second-site reversion linked to the original mutant by 20 map units and therefore most likely in a second gene. Note that auxotrophs equal half the recombinants.

19. **a.** A lack of revertants suggests either a deletion or an inversion within the gene.

b. To understand these data, recall that half the progeny should come from the wild-type parent.

Prototroph A: because 100% of the progeny are prototrophic, a reversion at the original mutant site may have occurred.

Prototroph B: half the progeny are parental prototrophs, and the remaining prototrophs, 28%, are the result of the new mutation. Notice that 28% is approximately equal to the 22% auxotrophs. The suggestion is that an unlinked suppressor mutation occurred, yielding independent assortment with the *nic* mutant.

Prototroph C: there are 496 "revertant" prototrophs (the other 500 are parental prototrophs) and 4 auxotrophs. This suggests that a suppressor mutation occurred in a site very close [$100\%(4 \times 2)/1000 = 0.8$ m.u.] to the original mutation.

20. **a.** To select for a nerve mutation that blocks flying, place *Drosophila* at the bottom of a cage and place a poisoned food source at the top of the cage.

b. Make antibodies against flagellar protein and expose mutagenized cultures to the antibodies.

c. Do filtration through membranes with variously sized pores.

d. Screen visually.

e. Go to a large shopping mall and set up a rotating polarized disk. Ask the passersby to look through the disk for a free evaluation of their vision and their need for sunglasses. People with normal vision will see light with a constant intensity through the disk. Those with polarized vision will see alternating dark and light.

f. Set up a Y tube (a tube with a fork giving the choice of two pathways) and observe whether the flies or unicellular algae move to the light or the dark pathway.

g. Set up replica cultures and expose one of the two plates to low doses of UV.

21. Yes. It will cause CG-to-TA transitions.

22. 15% are essential gene functions (such as enzymes required for DNA replication or protein synthesis).

25% are auxotrophs (enzymes required for the synthesis of amino acids or the metabolism of sugars, etc.).

60% are redundant or pathways not tested (genes for histones, tubulin, ribosomal RNAs, etc., are present in multiple copies; the yeast may require many genes under only unique or special situations or in other ways that are not necessary for life in the "lab").

23. The allele for NF must have arisen spontaneously in one of the parents' germ lines. Depending on when this mutation happened (the size of the mutant clone of germ-line tissue), their chance of having another affected child would range from 0 to 50% (the latter number if the entire germ line was mutant).

24. ***Unpacking the Problem***

a.

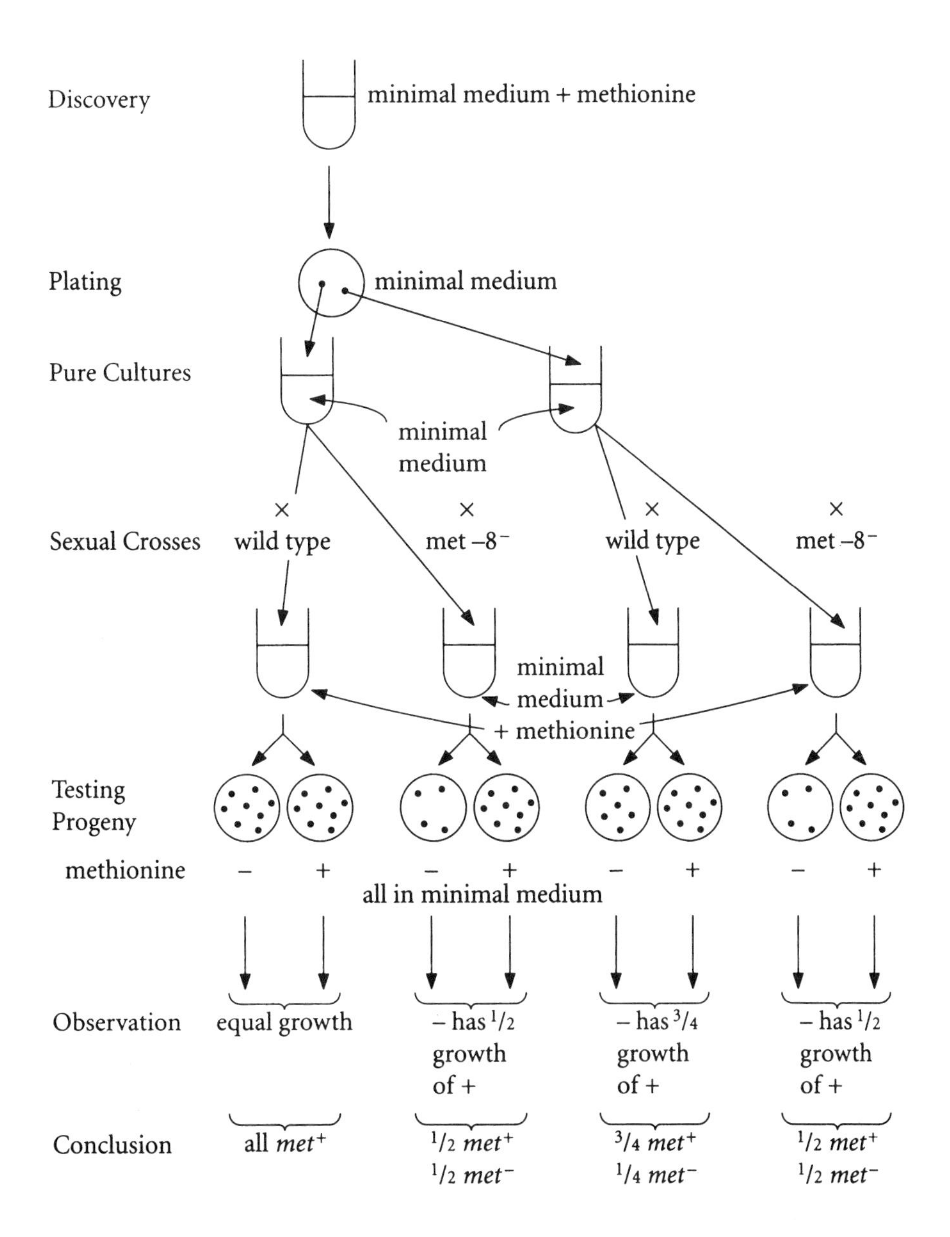

b. **Haploid** refers to possessing a single copy of the genome.

Auxotrophic means that an organism requires dietary provision of some substance that normally is not required by members of its species.

Methionine is an amino acid.

Asexual spores are a mode of propagation used by some species in which the spores are derived from an organism without a genetic contribution from another organism. Of necessity, the spores have the same number of genomes as the organism from which they are derived.

Prototrophic means that an organism does not have any special dietary requirements beyond those normal for the species.

A **colony** is a collection of cells or organisms all derived mitotically from a single cell or organism and all possessing the same genotype.

A **mutation** is the process that generates alternative forms of genes, and it results in an inherited difference between parent and progeny.

c. The "8" in *met-8* refers to the eighth locus found that leads to a methionine requirement. It is unnecessary to know the specifics of the mutation in order to work the problem.

d. The following crosses were made in this problem:

Cross 1: prototroph 1 × wild type
Cross 2: prototroph 1 × backcross to *met-8*
Cross 3: prototroph 2 × wild type
Cross 4: prototroph 2 × backcross to *met-8*

e. Use *met-8** to indicate the prototroph derived from the *met-8* strain.

Cross 1: *met-8** 1 × *met-8*$^+$
Cross 2: *met-8** 1 × *met-8*
Cross 3: *met-8** 2 × *met-8*$^+$
Cross 4: *met-8** 2 × *met-8*

f. In this organism, asexual spores give rise to an organism that is capable of forming sexual spores following a mating. Therefore, the original mutation occurred in somatic tissue that subsequently gave rise to germinal tissue.

g. Because the trait being selected is the ability to grow in the absence of methionine, a reversion is being studied.

h. Only two revertants were observed because reversion occurs at a much lower frequency than forward mutation.

i. The millions of asexual spores did not grow because they required methionine, and the medium used did not contain methionine.

j. A low percentage of the millions of spores that did not grow would be expected to have other mutations that rendered them incapable of growth. In addition, a low percentage would be expected to have chromosome abnormalities that would lead to death.

k. The wild type used in this experiment was prototrophic, by definition; that is, **wild type** refers to the norm for a species, which means "prototrophic."

l. It is highly unlikely that visual inspection could distinguish between wild type and prototrophic revertants.

m. One way to select for a *met-8* mutation is to grow a large number of spores on a medium that lacks methionine. Filtration will separate those spores capable of growth from those incapable of growth. Once spores have been isolated that are incapable of growth in a medium lacking methionine, they can be tested for a methionine mutation by plating them on medium containing methionine. If they are capable of growth on this second medium, they are methionine auxotrophs.

n. The starting auxotrophic spores were haploid. Both mitotic crossing-over and haploidization require diploids. Therefore, it is unlikely that either process is involved with producing the observed results.

o. Cross 1: *met-8* × wild type → 1 *met-8*:1 wild type
Cross 2: *met-8* × *met-8* → all *met-8*
Cross 3: wild type × wild type → all wild type

p. While the analysis could have been conducted using tetrad analysis, it is more likely that random selection of progeny was used.

q.

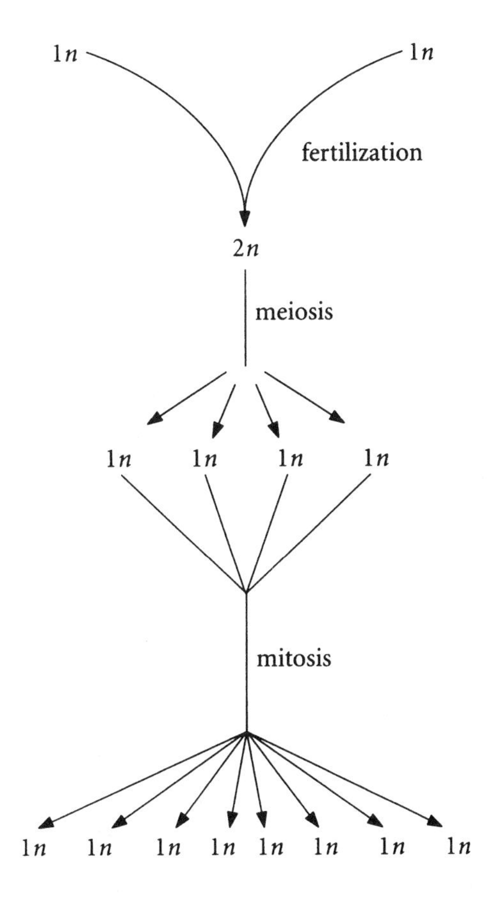

r. If a 3:1 ratio is obtained in haploids, then two genes must be segregating.

Solution to the Problem

a., b. The pattern of growth for prototroph 1 suggests that it is a reversion of the original mutation. When crossed with wild type, a reversion would be expected to produce all *met*$^+$ progeny, and when backcrossed, it would be expected to produce a 1:1 ratio. Let the reversion be symbolized by *met-8**. The crosses are:

	Cross 1		Cross 2
P	*met-8** × *met-8*$^+$	P	*met-8** × *met-8*
	↓		↓
F_1	$^1/_2$ *met-8** prototrophic	F_1	$^1/_2$ *met-8** prototrophic
	$^1/_2$ *met-8*$^+$ prototrophic		$^1/_2$ *met-8* auxotrophic

The pattern of growth for prototroph 2 suggests that a suppressor at another, unlinked locus is responsible for its prototrophic growth. Let *met-s* symbolize this suppressor. Then the crosses are

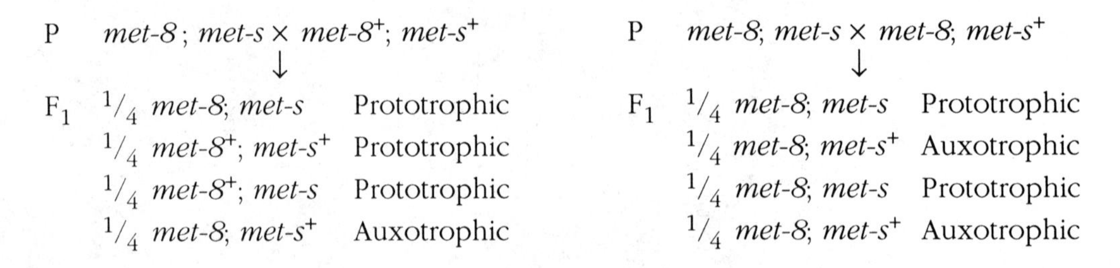

	Cross 1			Cross 2	
P	*met-8*; *met-s* × *met-8*$^+$; *met-s*$^+$		P	*met-8*; *met-s* × *met-8*; *met-s*$^+$	
	↓			↓	
F_1	$^1/_4$ *met-8*; *met-s*	Prototrophic	F_1	$^1/_4$ *met-8*; *met-s*	Prototrophic
	$^1/_4$ *met-8*$^+$; *met-s*$^+$	Prototrophic		$^1/_4$ *met-8*; *met-s*$^+$	Auxotrophic
	$^1/_4$ *met-8*$^+$; *met-s*	Prototrophic		$^1/_4$ *met-8*; *met-s*	Prototrophic
	$^1/_4$ *met-8*; *met-s*$^+$	Auxotrophic		$^1/_4$ *met-8*; *met-s*$^+$	Auxotrophic

8 Chromosome Mutations

1. **a.** Cytologically, deletions lead to shorter chromosomes with missing bands (if banded) and an unpaired loop during meiotic pairing when heterozygous. Genetically, deletions are usually lethal when homozygous, do not revert, and when heterozygous, lower recombinational frequencies and can result in "pseudodominance" (the expression of recessive alleles on one homolog that are deleted on the other). Occasionally, heterozygous deletions express an abnormal (mutant) phenotype.

   b. Cytologically, duplications lead to longer chromosomes and, depending on the type, unique pairing structures during meiosis when heterozygous. These may be simple unpaired loops or more complicated twisted loop structures. Genetically, duplications can lead to asymmetric pairing and unequal crossing-over events during meiosis, and duplications of some regions can produce specific mutant phenotypes.

   c. Cytologically, inversions can be detected by banding, and when heterozygous, they show the typical twisted "inversion" loop during homologous pairing. Pericentric inversions can result in a change in the p:q ratio. Genetically, no viable crossover products are seen from recombination within the inversion when heterozygous, and as a result, flanking genes show a decrease in RF.

   d. Cytologically, reciprocal translocations may be detected by banding, or they may drastically change the size of the involved chromosomes as well as the positions of their centromeres. Genetically, they establish new linkage relationships. When heterozygous, they show the typical cross structure during meiotic pairing and cause a diagnostic 50% reduction of viable gamete production, leading to semisterility.

2.

a. paracentric inversion

b. deletion

c. pericentric inversion

d. duplication

3. **a.** The products of crossing-over within the inversion will be inviable when the inversion is heterozygous. This paracentric inversion spans 25% of the region between the two loci and therefore will reduce the observed recombination between these genes by a similar percentage (i.e., 9%.) The observed RF will be 27%.

b. When the inversion is homozygous, the products of crossing-over within the inversion will be viable, so the observed RF will be 36%.

4. The following represents the crosses that are described in this problem:

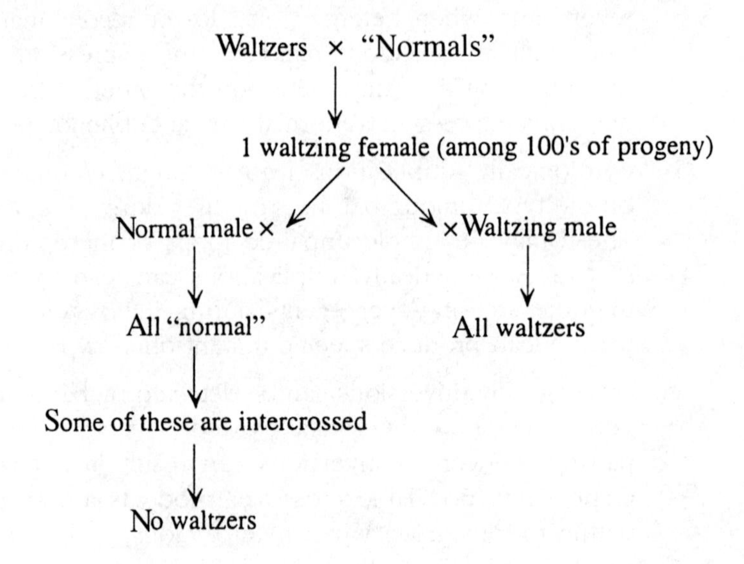

The single waltzing female that arose from a cross between waltzers and normals is expressing a recessive gene. It is possible that this represents a new "waltzer" mutation that was inherited from one of the "normal" mice, but given the cytological evidence (the presence of a shortened chromosome), it is more likely that this exceptional female inherited a deletion of the wild type allele, which allowed expression of the mutant recessive phenotype.

When this exceptional female was mated to a waltzing male, all the progeny were waltzers; when mated to a normal male, all the progeny were normal. When some of these normal offspring were intercrossed, there were no progeny that were waltzers. If a "new" recessive waltzer allele had been inherited, all these "normal" progeny would have been w^+/w. Any intercross should have therefore produced 25% waltzers. On the other hand, if a deletion had occurred, half the progeny would be w^+/w and half would be $w^+/w^{deletion}$. If $w^+/w^{deletion}$ are intercrossed, 25% of the progeny would not develop (the homozygous deletion would likely be lethal), and no waltzers would be observed. This is consistent with the data.

5. This problem uses a known set of overlapping deletions to order a set of mutants. This is called **deletion mapping** and is based on the expression of the recessive mutant phenotype when heterozygous with a deletion of the corresponding allele on the other homolog. For example, mutants *a*, *b*, and *c* are all expressed when heterozygous with Del1. Thus it can be assumed that these genes are deleted in Del1. When these results are compared with the crosses with Del2 and it is discovered that these progeny are b^+, the location of gene *b* is mapped to the region deleted in Del1 that is not deleted in Del2. This logic can be applied in the following way:

Compare deletions 1 and 2: this places allele *b* more to the left than alleles *a* and *c*. The order is *b* (*a*, *c*), where the parentheses indicate that the order is unknown.

Compare deletions 2 and 3: this places allele *e* more to the right than (*a*, *c*). The order is *b* (*a*, *c*) *e*.

Compare deletions 3 and 4: allele *a* is more to the left than *c* and *e*, and *d* is more to the right than *e*. The order is *b a c e d*.

Compare deletions 4 and 5: allele *f* is more to the right than *d*. The order is *b a c e d f*.

Allele	Band
b	1
a	2
c	3
e	4
d	5
f	6

6. The data suggest that one or both breakpoints of the inversion are located within an essential gene, causing a recessive lethal mutation.

7. a. Single crossovers between a gene and its centromere lead to a tetratype (second-division segregation). Thus a total of 20% of the asci should show second-division segregation, and 80% will show first-division segregation. The following are representative asci:

$un3^{+}\ ad3^{+}$	$un3^{+}\ ad3^{+}$	$un3^{+}\ ad3^{+}$	$un3^{+}\ ad3$	$un3^{+}\ ad3$
$un3^{+}\ ad3^{+}$	$un3^{+}\ ad3^{+}$	$un3^{+}\ ad3^{+}$	$un3^{+}\ ad3$	$un3^{+}\ ad3$
$un3^{+}\ ad3^{+}$	$un3^{+}\ ad3$	$un3^{+}\ ad3$	$un3^{+}\ ad3^{+}$	$un3^{+}\ ad3^{+}$
$un3^{+}\ ad3^{+}$	$un3^{+}\ ad3$	$un3^{+}\ ad3$	$un3^{+}\ ad3^{+}$	$un3^{+}\ ad3^{+}$
$un3\ ad3$	$un3\ ad3^{+}$	$un3\ ad3$	$un3\ ad3^{+}$	$un3\ ad3$
$un3\ ad3$	$un3\ ad3^{+}$	$un3\ ad3$	$un3\ ad3^{+}$	$un3\ ad3$
$un3\ ad3$	$un3\ ad3$	$un3\ ad3^{+}$	$un3\ ad3$	$un3\ ad3^{+}$
$un3\ ad3$	$un3\ ad3$	$un3\ ad3^{+}$	$un3\ ad3$	$un3\ ad3^{+}$
80%	5%	5%	5%	5%

In all cases, the "upside-down" version would be equally likely.

b. The aborted spores could result from a crossing-over event within an inversion of the wild type compared with the standard strain. Crossing-over within heterozygous inversions leads to unbalanced chromosomes and nonviable spores. This could be tested by using the wild type from Hawaii in mapping experiments of other markers on chromosome 1 in crosses with the standard strain and looking for altered map distances.

8. a. The Sumatra chromosome contains a pericentric inversion when compared with the Borneo chromosome.

b.

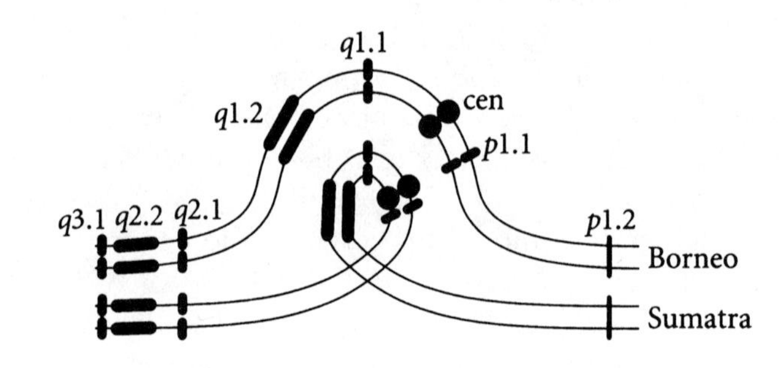

c.

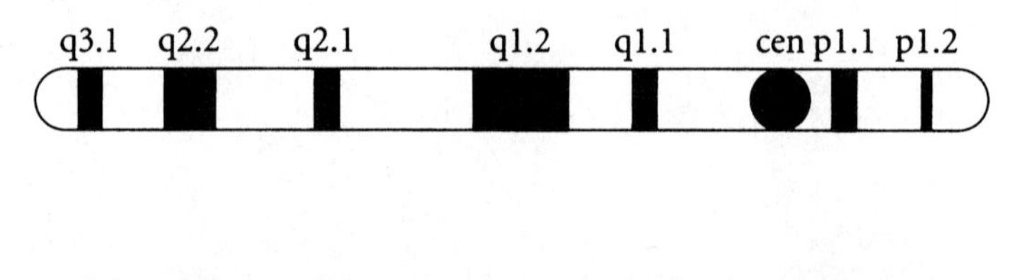

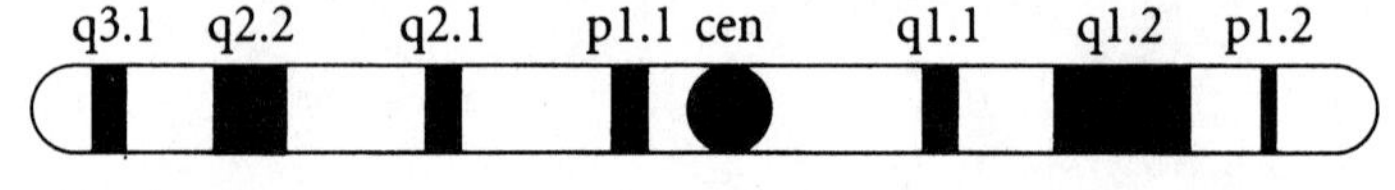

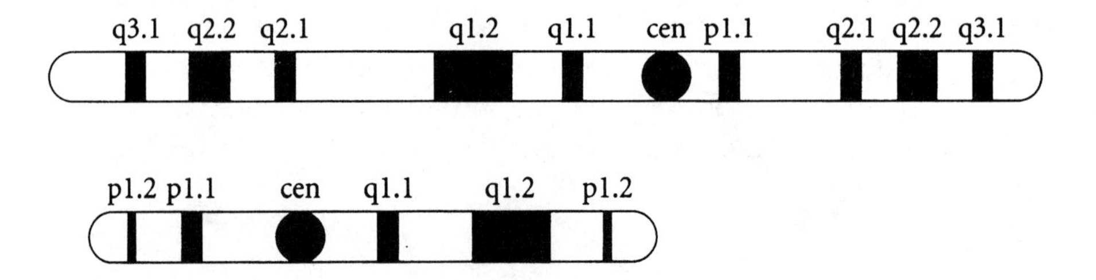

d. Recall that all single crossovers within the inverted region will lead to four meiotic products: two that will be viable, nonrecombinant (parental) types and two that will be extremely unbalanced (most likely nonviable), recombinant types. In other words, if 30% of the meioses have a crossover in this region, 15% of the gametes will not lead to viable progeny. That means that 85% of the gametes should produce viable progeny.

9. ***Unpacking the Problem***

a. A "gene for tassel length" means that there is a gene with at least two alleles (*T* and *t*) that controls the length of the tassel. A "gene for rust resistance" means that there is a gene that determines whether the corn plant is resistant to a rust infection or not (*R* and *r*).

b. The precise meaning of the allelic symbols for the two genes is irrelevant to solving the problem because what is being investigated is the distance between the two genes.

c. A **locus** is the specific position occupied by a gene on a chromosome. It is implied that gene loci are the same on both homologous chromosomes. The gene pair can consist of identical or different alleles.

d. Evidence that the two genes are normally on separate chromosomes would have come from previous experiments showing that the two genes independently assort during meiosis.

e. Routine crosses could consist of F_1 crosses, F_2 crosses, backcrosses, and testcrosses.

f. The genotype *T/t*; *R/r* is a double heterozygote, or dihybrid, or F_1 genotype.

g. The pollen parent is the "male" parent that contributes to the pollen tube nucleus, the endosperm nucleus, and the progeny.

h. Testcrosses are crosses that involve a genotypically unknown and a homozygous recessive organism. They are used to reveal the complete genotype of the unknown organism and to study recombination during meiosis.

i. The breeder was expecting to observe 1 *T/t*; *R/r*:1 *T/t*; *r/r*:1 *t/t*; *R/r*:1 *t/t*; *r/r*.

j. Instead of a 1:1:1:1 ratio indicating independent assortment, the testcross indicated that the two genes were linked, with a genetic distance of 100%(3 + 5)/210 = 3.8 map units.

k. The equality and predominance of the first two classes indicate that the parentals were *TR/t r*.

l. The equality and lack of predominance of the second two classes indicate that they represent recombinants.

m. The gametes leading to this observation were:

46.7% *TR*	1.4% *Tr*
49.5% *t r*	2.4% *tR*

n. 46.7% *TR*
49.5% *t r*

o. 1.4% *Tr*
2.4% *tR*

p. *Tr* and *tR*

q. *T* and *R* are linked, as are *t* and *r*.

r. Two genes on separate chromosomes can become linked through a translocation.

s. One parent of the hybrid plant contained a translocation that linked the *T* and *R* alleles and the *t* and *r* alleles.

t. A corn cob is a structure that holds on its surface the progeny of the next generation.

u.

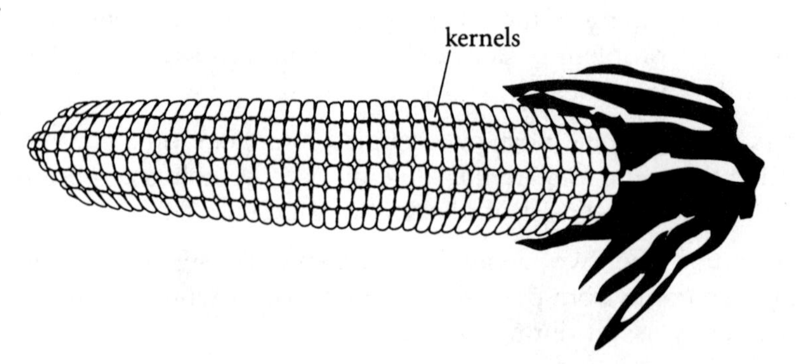

v.

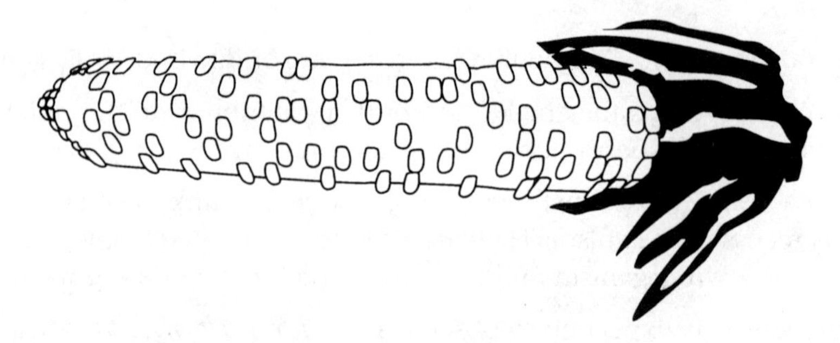

w. A kernel is one progeny on a corn cob.

x. Absence of half the kernels, or 50% aborted progeny (semisterility), could result from the random segregation of one normal with one translocated chromosome (T1 + N2 and T2 + N1) during meioses in a parent that is heterozygous for a reciprocal translocation.

y. Approximately 50% of the progeny died. It was the "female" that was heterozygous for the translocation.

Solution to the Problem

a. The progeny are not in the 1:1:1:1 ratio expected for independent assortment; instead, the data indicate close linkage. And, half the progeny did not develop, indicating semisterility.

b. These observations are best explained by a translocation that brought the two loci close together.

c. Parents: $TR/tr \times t/t;\ r/r$

Progeny:

98	$TR/t;\ r$
104	$tr/t;\ r$
3	$Tr/t;\ r$
5	$tR/t;\ r$

d. Assume a translocation heterozygote in coupling. If pairing is as diagrammed below, then you would observe the following:

No Crossover

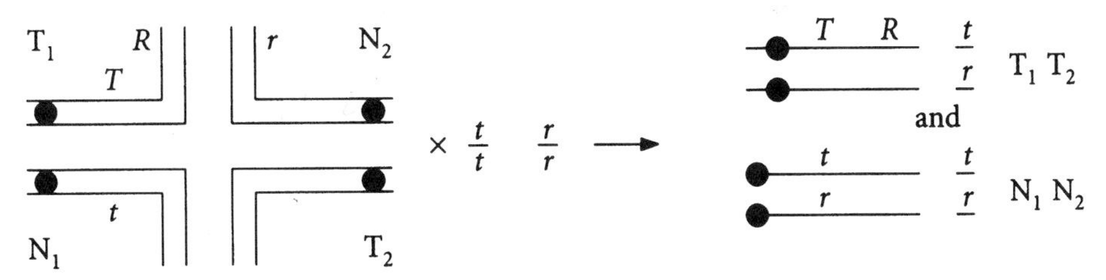

Crossover between T and R

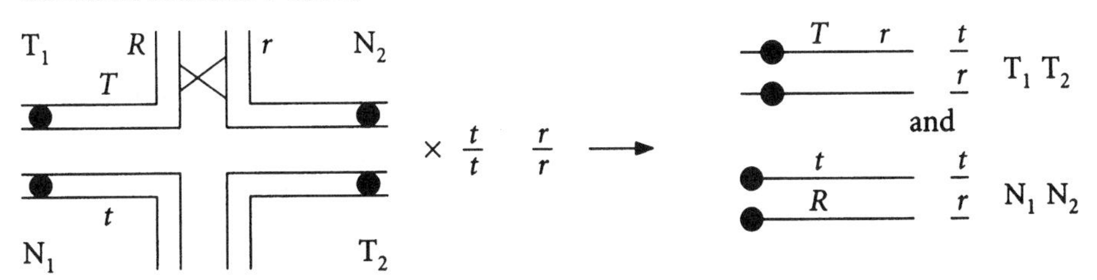

e. The two recombinant classes result from a recombination event followed by proper segregation of chromosomes, as diagrammed above.

10. The cross was

P X^{e+}/Y (irradiated) $\times$ X^{e}/X^{e}

F_1 Most X^{e}/Y yellow males

Two ? gray males

a. Gray male 1 was crossed with a yellow female, yielding yellow females and gray males, which is reversed sex linkage. If the e^+ allele was translocated to the Y chromosome, the gray male would be X^e/Y^{e+} or gray. When crossed with yellow females, the results would be

X^e/Y^{e+} Gray males

X^e/Xe Yellow females

b. Gray male 2 was crossed with a yellow female, yielding gray and yellow males and females in equal proportions. If the e^+ allele was translocated to an autosome, the progeny would be as below, where "A" indicates autosome:

P $A^{e+}/A;\ X^e/Y \times A/A;\ X^e/X^e$

F_1

$A^{e+}/A;\ X^e/X^e$	Gray female
$A^{e+}/A;\ X^e/Y$	Gray male
$A/A;\ X^e/X^e$	Yellow female
$A/A;\ X^e/Y$	Yellow male

11. Cross 1: independent assortment of the 2 genes (expected for genes on separate chromosomes).
Cross 2: the 2 genes now appear to be linked (the observed RF is 1%); also, half the progeny are inviable. These data suggest a reciprocal translocation occurred and both genes are very close to the breakpoints.
Cross 3: the viable spores are of 2 types: half contain the normal (non-translocated chromosomes) and half contain the translocated chromosomes.

12. The break point can be treated as a gene with two "alleles," one for normal fertility and one for semisterility. The problem thus becomes a two-point cross.

Parentals	764	Semisterile *Pr*
	727	Normal *pr*
Recombinants	145	Semisterile *pr*
	186	Normal *Pr*
	1822	

100%(145 + 186)/1822 = 18.17 m.u.

13. The percent degeneration seen in the progeny of the exceptional rat is roughly 50% larger than that seen in the progeny from the normal male. Semisterility is an important diagnostic for translocation heterozygotes. This could be verified by cytological observation of the meiotic cells from the exceptional male.

14.

Klinefelter syndrome	XXY male
Down syndrome	Trisomy 21
Turner syndrome	XO female

15. Create a hybrid by crossing the two plants and then double the chromosomes with a treatment that disrupts mitosis such as colchicine treatment. Alternatively, diploid somatic cells from the two plants could be fused and then grown into plants through various culture techniques.

16. a. $a^+/a^+/a/a \times a/a/a/a$

$\downarrow$ $\downarrow$

Gametes: $^1/_6$ a^+/a^+ ; a/a
$^2/_3$ a^+/a
$^1/_6$ a/a

Among the progeny of this cross, the phenotypic ratio will be 5 wild-type (a^+):1 a.

b. $a^+/a/a/a \times a/a/a/a$

$\downarrow$ $\downarrow$

Gametes: $^1/_2$ a^+/a ; a/a
$^1/_2$ a/a

Among the progeny of this cross, the phenotypic ratio will be 1 wild-type (a^+):1 a.

c. $a^+/a/a/a \times a^+/a/a/a$

$\downarrow$ $\downarrow$

Gametes: $^1/_2$ a^+/a ; $^1/_2$ a^+/a
$^1/_2$ a/a ; $^1/_2$ a/a

Among the progeny of this cross, the phenotypic ratio will be 3 wild-type (a^+):1 a.

d. $a^+/a^+/a/a \times a^+/a/a/a$

$\downarrow \qquad \downarrow$

Gametes: $^1/_6$ a^+/a^+ $\qquad$ $^1/_2$ a^+/a
$^2/_3$ a^+/a $\qquad$ $^1/_2$ a/a
$^1/_6$ a/a

Among the progeny of this cross, the phenotypic ratio will be 11 wild-type (a^+):1 a.

17. a. $b^+/b/b \times b/b$

$\downarrow \qquad \downarrow$

Gametes: $^1/_6$ b^+ $\qquad$ b
$^1/_3$ b
$^1/_3$ b^+/b
$^1/_6$ b/b

Among the progeny of this cross, the phenotypic ratio will be 1 wild-type (b^+):1 b.

b. $b^+/b^+/b \times b/b$

$\downarrow \qquad \downarrow$

Gametes: $^1/_6$ b $\qquad$ b
$^1/_3$ b^+
$^1/_3$ b^+/b
$^1/_6$ b^+/b^+

Among the progeny of this cross, the phenotypic ratio will be 5 wild-type (b^+):1 b.

c. $b^+/b^+/b \times b^+/b$

$\downarrow \qquad \downarrow$

Gametes: $^1/_6$ b $\qquad$ $^1/_2$ b
$^1/_3$ b^+ $\qquad$ $^1/_2$ b^+
$^1/_3$ b^+/b
$^1/_6$ b^+/b^+

Among the progeny of this cross, the phenotypic ratio will be 11 wild-type (b^+):1 b.

18. a., b., c. One of the parents of the woman with Turner syndrome (XO) must have been a carrier for colorblindness, an X-linked recessive disorder. Because her father has normal vision, she could not have obtained her only X from him. Therefore, nondisjunction occurred in her father. A sperm lacking a sex chromosome fertilized an egg with the X chromosome carrying the colorblindness allele. The nondisjunctive event could have occurred during either meiotic division.

d. If the colorblind patient had Klinefelter syndrome (XXY), then both X's must carry the allele for colorblindness. Therefore, nondisjunction had to occur in the mother. Remember that during meiosis I, given no crossover between the gene and the centromere, allelic alternatives separate from each other. During meiosis II, identical alleles on sister chromatids separate. Therefore, the nondisjunctive event had to occur during meiosis II because both alleles are identical.

19. a. If a 6x were crossed with a 4x, the result would be 5x.

b. Cross *A/A* with *a/a/a/a* to obtain *A/a/a.*

c. The easiest way is to expose the *A/a** plant cells to colchicine for one cell division. This will result in a doubling of chromosomes to yield *A/A/a*/a**.

d. Cross 6x (*a/a/a/a/a/a*) with 2x (*A/A*) to obtain *A/a/a/a.*

e. In culture, expose haploid plants cells to the herbicide and select for resistant colonies. Treat cells that grow with colchicine to obtain diploid cells.

20. Consider the following table, in which "L" and "S" stand for 13 large and 13 small chromosomes, respectively:

Hybrid	Chromosomes
G. hirsutum × *G. thurberi*	S, S, L
G. hirsutum × *G. herbaceum*	S, L, L
G. thurberi × *G. herbaceum*	S, L

Each parent in the cross must contribute half its chromosomes to the hybrid offspring. It is known that *G. hirsutum* has twice as many chromosomes as the other two species. Furthermore, its chromosomes are composed of chromosomes donated by the other two species. Therefore, the genome of *G. hirsutum* must consist of one large and one small set of chromosomes. Once this is realized, the rest of the problem essentially solves itself. In the first hybrid, the genome of *G. thurberi* must consist of one set of small chromosomes. In the second hybrid, the genome of *G. herbaceum* must consist of one set of large chromosomes. The third hybrid confirms the conclusions reached from the first two hybrids.

The original parents must have had the following chromosome constitution:

G. hirsutum	26 large, 26 small
G. thurberi	26 small
G. herbaceum	26 large

G. hirsutum is a polyploid derivative of a cross between the two Old World species. This could easily be checked by looking at the chromosomes.

21. a. Loss of one X in the developing fetus after the two-cell stage

b. Nondisjunction leading to Klinefelter syndrome (XXY), followed by a nondisjunctive event in one cell for the Y chromosome after the two-cell stage, resulting in XX and XXYY

c. Nondisjunction of the X at the one-cell stage

d. Fused XX and XY zygotes (from the separate fertilizations either of two eggs or of an egg and a polar body by one X-bearing and one Y-bearing sperm)

e. Nondisjunction of the X at the two-cell stage or later

22. Type a: the extra chromosome must be from the mother. Because the chromosomes are identical, nondisjunction had to have occurred at M_{II}.
Type b: the extra chromosome must be from the mother. Because the chromosomes are not identical, nondisjunction had to have occurred at M_I.
Type c: the mother correctly contributed one chromosome, but the father did not contribute any chromosome 4. Therefore, nondisjunction occurred in the male during either meiotic division.

23. Cross 1: P $b\,e^+/b\,e^+ \times b^+\,e/b^+\,e$
F_1 $b^+\,e/b\,e^+$

Cross 2: P X/X ; $b^+\,e/b\,e^+ \times$ X/Y ; $b\,e/b\,e$
F_1 Expect 1 $b\,e^+/b\,e$:1 $b^+\,e/b\,e$, X/X and X/Y
one rare observed X/X ; $b^+\,e^+$

a. The common progeny are $b^+\,e/b\,e$ and $b\,e^+/b\,e$.

b. The rare female could have come from crossing-over, which would have resulted in a gamete that was $b^+\,e^+$. The rare female also could have come from nondisjunction that gave a gamete that was $b\,e^+/b^+\,e$. Such a gamete might give rise to viable progeny.

c. If the female had been wild type ($b^+\,e^+/b\,e$) as a result of crossing-over, her progeny would have been as follows:

Parental:	$b^+\,e^+/b\,e$	Wild type (common)
	$b\,e/b\,e$	Bent, eyeless (common)
Recombinant:	$b\,e^+/b\,e$	Bent (rare)
	$b^+\,e/b\,e$	Eyeless (rare)

These expected results are very far from what was observed, so the rare female was not the result of recombination.

If the female had been the product of nondisjunction ($b\,e^+/b^+\,e/b\,e$), her progeny when crossed to $b\,e/b\,e$ would be as follows:

$^1/_6$	$b^+\,e/b\,e$	Eyeless
$^1/_6$	$b\,e^+/b\,e/b\,e$	Bent
$^1/_6$	$b^+\,e/b\,e/b\,e$	Eyeless
$^1/_6$	$b\,e^+/b\,e$	Bent
$^1/_6$	$b\,e/b\,e$	Bent, eyeless
$^1/_6$	$b\,e^+/b^+\,e/b\,e$	Wild type

Overall, 2 bent:2 eyeless:1 bent eyeless:1 wild type

These results are in accord with the observed results, indicating that the female was a product of nondisjunction.

24. a. The cross is *P/P/p* × *p/p*.

The gametes from the trisomic parent will occur in the following proportions:

$^1/_6$ *p*
$^2/_6$ *P*
$^1/_6$ *P/P*
$^2/_6$ *P/p*

Only gametes that are *p* can give rise to potato leaves because potato is recessive. Therefore, the ratio of normal to potato will be 5:1.

b. If the gene is not on chromosome 6, there should be a 1:1 ratio of normal to potato.

25. The generalized cross is *A/A/A* × *a/a*, from which *A/A/a* progeny were selected. These progeny were crossed with *a/a* individuals, yielding the results given. Assume for a moment that each allele can be distinguished from the other, and let 1 = *A*, 2 = *A*, and 3 = *a*. The gametic combinations possible are

1-2 (*A/A*) and 3 (*a*)
1-3 (*A/a*) and 2 (*A*)
2-3 (*A/a*) and 1 (*A*)

Because only diploid progeny were examined in the cross with *a/a*, the progeny ratio should be 2 wild type:1 mutant if the gene is on the trisomic chromosome. With this in mind, the table indicates that *y* is on chromosome 1, *cot* is on chromosome 7, and *h* is on chromosome 10. Genes *d* and *c* do not map to any of these chromosomes.

26. Recall that ascospores are haploid. The normal genotype associated with the phenotype of each spore is given below:

1	2	3
$b^+ f^+$	$b f^+$	$b^+ f$
$b^+ f^+$	$b f^+$	$b^+ f$
$b^+ f^+$	abort	$b^+ f^+$
$b^+ f^+$	abort	$b^+ f^+$
abort	$b^+ f$	$b f$
abort	$b^+ f$	$b f$
abort	$b^+ f$	$b f^+$
abort	$b^+ f$	$b f^+$

a. For the first ascus, the most reasonable explanation is that nondisjunction occurred at the first meiotic division. Second-division nondisjunction or chromosome loss are two explanations of the second ascus. Crossing-over best explains the third ascus.

b.

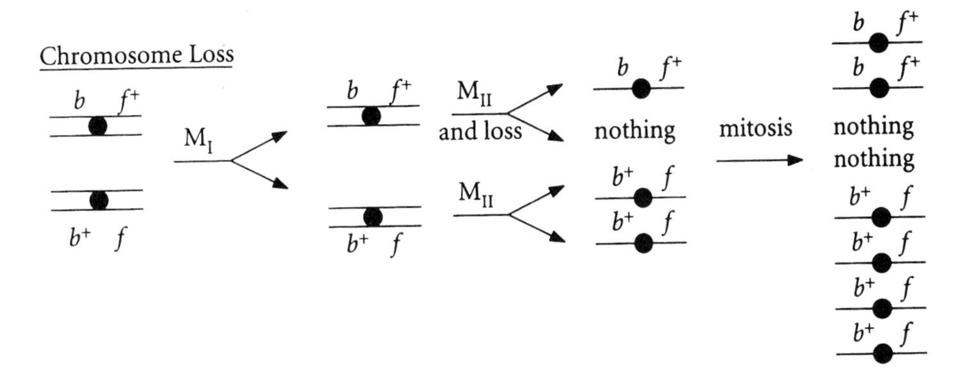
Chromosome Loss
b f+
b+ f
MI
MII
and loss
MII
nothing
mitosis
nothing
nothing

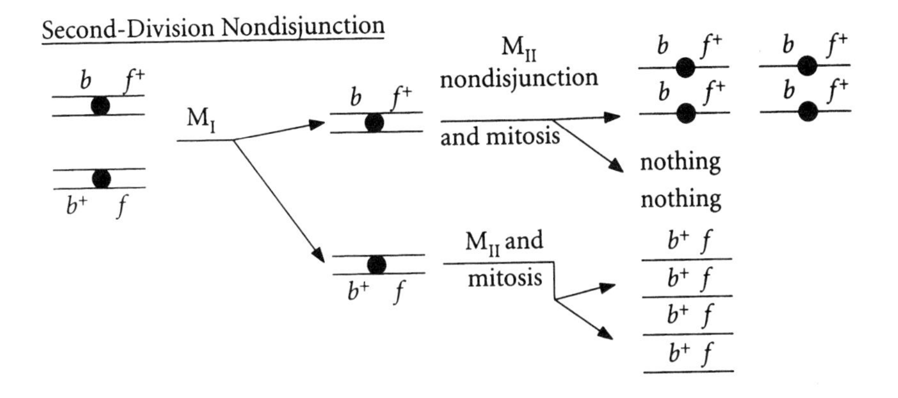
Second-Division Nondisjunction
b f+
b+ f
MI
MII
nondisjunction
and mitosis
nothing
nothing
MII and
mitosis

9 The Genetics of Bacteria and Phages

1. An Hfr strain has the fertility factor F integrated into the chromosome. An F^+ strain has the fertility factor free in the cytoplasm. An F^- strain lacks the fertility factor.

2. All cultures of F^+ strains have a small proportion of cells in which the F factor is integrated into the bacterial chromosome and are, by definition, Hfr cells. These Hfr cells transfer markers from the host chromosome to a recipient during conjugation.

3. **a.** Hfr cells involved in conjugation transfer host genes in a linear fashion. The genes transferred depend on both the Hfr strain and the length of time during which the transfer occurred. Therefore, a population containing several different Hfr strains will appear to have an almost random transfer of host genes. This is similar to generalized transduction, in which the viral protein coat forms around a specific amount of DNA rather than specific genes. In generalized transduction, any gene can be transferred.

   b. F' factors arise from improper excision of an Hfr from the bacterial chromosome. They can have only specific bacterial genes on them because the integration site is fixed for each strain. Specialized transduction resembles this in that the viral particle integrates into a specific region of the bacterial chromosome and then, upon improper excision, can take with it only specific bacterial genes. In both cases, the transferred gene exists as a second copy.

4. Generalized transduction occurs with lytic phages that enter a bacterial cell, fragment the bacterial chromosome, and then, while new viral particles are

being assembled, improperly incorporate some bacterial DNA within the viral protein coat. Because the amount of DNA, not the information content of the DNA, is what governs viral particle formation, any bacterial gene can be included within the newly formed virus. In contrast, specialized transduction occurs with improper excision of viral DNA from the host chromosome in lysogenic phages. Because the integration site is fixed, only those bacterial genes very close to the integration site will be included in a newly formed virus.

5. While the interrupted-mating experiments will yield the gene order, it will be relative only to fairly distant markers. Thus, the precise location cannot be pinpointed with this technique. Generalized transduction will yield information with regard to very close markers, which makes it a poor choice for the initial experiments because of the massive amount of screening that would have to be done. Together, the two techniques allow, first, for a localization of the mutant (interrupted-mating) and, second, for precise determination of the location of the mutant (generalized transduction) within the general region.

6. This problem is analogous to forming long gene maps with a series of three-point testcrosses. Arrange the four sequences so that their regions of overlap are aligned:

$$\overline{M}\text{—}Z\text{—}X\text{—}W\text{—}C$$
$$W\text{—}C\text{—}N\text{—}A\text{—}L$$
$$A\text{—}L\text{—}B\text{—}R\text{—}U$$
$$B\text{—}R\text{—}U\text{—}\underline{M\text{—}Z}$$

The regions with the bars above or below are identical in sequence (and "close" the circular chromosome). The correct order of markers on this circular map is

—M—Z—X—W—C—N—A—L—B—R—U—

7. To interpret the data, the following results are expected:

Cross	Result
$F^+ \times F^-$	(L) low number of recombinants
Hfr $\times$ F^-	(M) many recombinants
Hfr $\times$ Hfr	(0) no recombinants
Hfr $\times$ F^+	(0) no recombinants
$F^+ \times F^+$	(0) no recombinants
$F^- \times F^-$	(0) no recombinants

The only strains that show both the (L) and the (M) result when crossed are 2, 3, and 7. These must be F^- since that is the only cell type that can participate in a cross and give either recombination result. Hfr strains will result in only (M) or (0), and F^+ will result in only (L) or (0) when crossed. Thus, strains 1 and 8 are F^+, and strains 4, 5, and 6 are Hfr.

8. Prototrophic strains of *E. coli* will grow on minimal media while auxotrophic strains will only grow on media supplemented with the required molecule(s). Thus, strain 3 is prototrophic (wild type), strain 4 is *met*$^-$, strain 1 is *arg*$^-$, and strain 2 is *arg*$^-$ *met*$^-$.

9. a.

Agar type	Selected genes
1	c^+
2	a^+
3	b^+

b. The order of genes is revealed in the sequence of colony appearance. Because colonies first appear on agar type 1, which selects for c^+, *c* must be first. Colonies next appear on agar type 3, which selects for b^+, indicating that *b* follows *c*. Allele a^+ appears last. The gene order is *c b a*. The three genes are roughly equally spaced.

c. In this problem you are looking at the cotransfer of the selected gene with the d^- allele (both from the Hfr). Cells that are d^- do not grow because the medium is lacking D and selects for those cells that are d^+. Therefore, the farther a gene is from gene *d*, the less likely cotransfer of the selected gene will occur with d^- and the more likely that colonies will grow (remain d^+). From the data, *d* is closest to *b* (only $^8/_{100}$ did not cotransfer d^- with b^+). It is also closer to *a* than it is to *c*. Thus the gene order is *c b d a* (or *a d b c*).

d. With no A or B in the agar, the medium selects for a^+ b^+, and the first colonies should appear at about 17.5 minutes.

10. ***Unpacking the Problem***

a. *E. coli* is a bacterium and a prokaryote.

b. *E. coli* can be grown in suspension or on an agar medium. The latter method allows for the identification of individual colonies, each a clone of descendants from a single cell (and visible to the naked eye when it reaches more than 10^7 cells).

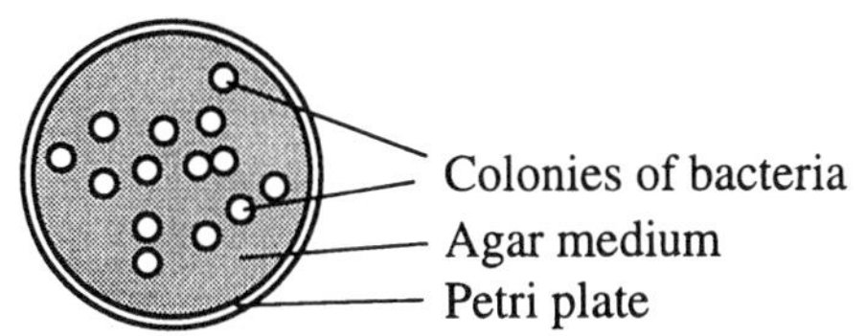

c. Naturally, *E. coli* is an enteric bacterium living symbiotically within the gut of host organisms (like us).

d. Minimal medium consists of inorganic salts, a carbon source for energy, and water.

e. Prototroph refers to the wild-type phenotype, or, in other words, an organism that can grow on minimal media. Auxotroph refers to a mutant that can grow only on a medium supplemented with one or more specific nutrients not required by the wild-type strain.

f. In this experiment, the Hfr and the exconjugants that can grow on minimal medium are prototrophs, while the recipient F^- and the exconjugants that do not grow on minimal medium are auxotrophs.

g. Unknown strains would be grown as individual colonies on medium enriched with proline and thiamine, and then cells from each colony could be picked (by a sterile toothpick, for example) and placed individually onto medium supplemented with either thiamine or proline or onto minimal medium. Proline and thiamine auxotrophs would be identified on the basis of growth patterns. For example, a *pro⁻* strain will grow only on medium supplemented with proline.

Instead of the labor-intensive method of individually picking cells, replica plating can be used to transfer some cells of each colony from a master plate (supplemented with proline and thiamine) to plates that contain the various media described above. The physical arrangement (and positional patterns) of colonies is used to identify the various colonies as they are transferred from plate to plate.

h. Proline is an amino acid and thiamine is a B_1 vitamin. Their chemical nature does not matter to the experiment other than that they are necessary chemicals for cell growth that prototrophs can synthesize from ingredients in minimal medium and specific auxotrophic mutants cannot.

i.

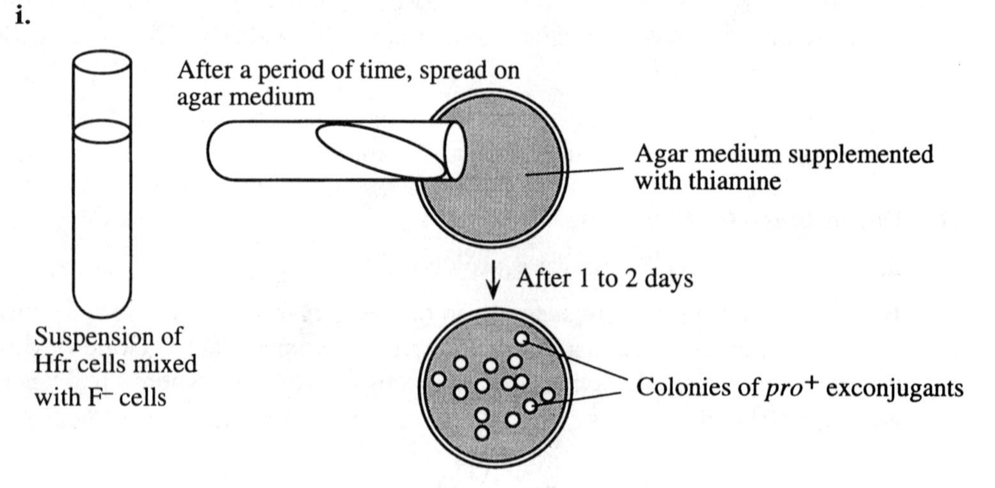

j. Interrupted-mating experiments are used to roughly map genes onto the circular bacterial chromosome.

k. The Hfr and F⁻ strains are mixed together in solution, and then at various times, samples are removed and put into a kitchen blender, vortexed (blender is turned on) for a few seconds to disrupt conjugation, and then plated onto a medium containing the appropriate supplements. The amount of time that has passed from the mixing of the strains to mating disruption is used as a measurement for mapping. The time of first appearance of a specific gene from the Hfr in the F⁻ cell gives the gene's relative position in minutes. Typically, the F⁻ cells are streptomycin resistant and the Hfr cells are streptomycin sensitive. The antibiotic is used in the various media to kill the Hfr cells (which are otherwise prototrophic) and allow only those F⁻ cells that have received the appropriate gene or genes from the Hfr to grow. In this case, it would be discovered that some of the F⁻ cells would become *thi*⁺ in samples taken earlier in the experiment than samples taken when they first become *pro*⁺.

l. In this experiment, there is no attempt to disrupt conjugation. The two strains are mixed and at some later (unspecified) time, plated onto medium containing thiamine. This selects for strains that are *pro*$^+$, since proline is not present in this medium.

m. Exconjugants are recipient cells (F^-) that now contain alleles from the donor (Hfr). Typically, the F^- cells are streptomycin resistant and the Hfr cells are streptomycin sensitive. The antibiotic is used in the various media to kill the Hfr cells and allow only the appropriate F^- exconjugants to grow.

n. The statement "*pro* enters after *thi*" is one of gene position and order relative to the transfer of the bacterial chromosome by a particular Hfr. For the Hfr in this experiment, transfer occurs such that the *pro* gene is transferred after the *thi* gene. Since this Hfr is also *pro*$^+$, it is this specific allele that is entering.

o. In this experiment, "fully supplemented" medium contains proline and thiamine.

p. All exconjugants are *pro*$^+$, since that is the way they were selected. Thus, those that do not grow on minimal medium must require thiamine.

q. Genetic exchange in prokaryotes does not take place between two whole genomes as it does in eukaryotes. It takes place between one complete circular genome, the F^-, and an incomplete linear genomic fragment donated by the Hfr. In this way, exchange of genetic information is nonreciprocal (from Hfr to F^-). Only even numbers of crossovers are allowed between the two DNAs, since the circular chromosome would become linear otherwise. This results in unidirectional exchange, since part of the DNA of the recipient chromosome is replaced by the DNA of the donor, while the other product (the rest of the donor DNA now with some recombined recipient DNA) is nonviable and lost.

r. In this experiment, the map distance is calculated by selecting for the last marker to enter (in this case *pro*$^+$) and then determining how often the earlier unselected marker (in this case *thi*$^+$) is also present. Look at the following diagram:

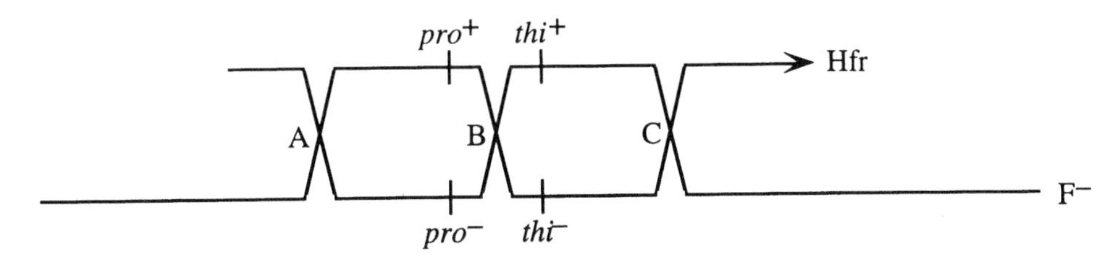

For the F^- cell to become *pro*$^+$, two recombination events have to occur—one in the region to the left (marked A) and one in either region to the right (marked B or C). Thus the percentage of *pro*$^+$ (second recombination within either B or C) that are *thi*$^-$ (second event only within region B) can be used to determine map distance, where 1% = 1 map unit.

Solution to the Problem

a. The two genotypes being cultured are pro^+ thi^- (grows only on media supplemented with thiamine) and pro^+ thi^+ (grows on minimal media).

b. Two recombination events must occur, one on either side of *pro* (since exconjugants were plated on medium supplemented with thiamine, only pro^+ cells would have grown). The pro^+ thi^- strains would have had recombination in regions labeled A and B, and the pro^+ thi^+ strains would have had recombination in regions labeled A and C.

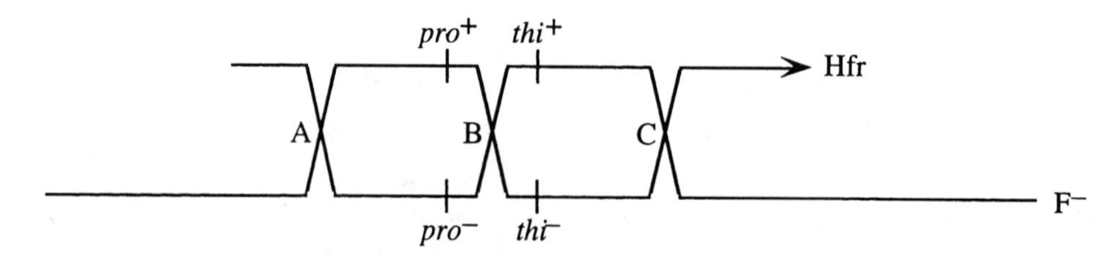

c. The distance between

$$pro \text{ and } thi = \frac{100\%(\text{the number of colonies that are } pro^+\ thi^-)}{\text{total number of } pro^+ \text{ colonies}}$$
$$= 100\%(40)/360 = 11.1\%$$

11. a. Determine the gene order by comparing arg^+ bio^+ leu^- with arg^+ bio^- leu^+. If the order were *arg leu bio*, four crossovers would be required to get arg^+ leu^- bio^+, while only two would be required to get arg^+ leu^+ bio^-. If the order is *arg bio leu*, four crossovers would be required to get arg^+ bio^- leu^+, and only two would be required to get arg^+ bio^+ leu^-. There are 8 recombinants that are arg^+ bio^+ leu^- and none that are arg^+ bio^- leu^+. On the basis of the frequencies of these two classes, the gene order is *arg bio leu.*

b. The *arg–bio* distance is determined by calculating the percentage of the exconjugants that are arg^+ bio^- leu^-. These cells would have had a crossing-over event between the *arg* and *bio* genes.

$$RF = 100\%(48)/376 = 12.76 \text{ m.u.}$$

Similarly, the *bio–leu* distance is estimated by the arg^+ bio^+ leu^- colony type.

$$RF = 100\%(8)/376 = 2.12 \text{ m.u.}$$

12. The most straightforward way would be to pick two Hfr strains that are near the genes in question but are oriented in opposite directions. Then, measure the time of transfer between two specific genes, in one case when they are transferred early and in the other when they are transferred late. For example,

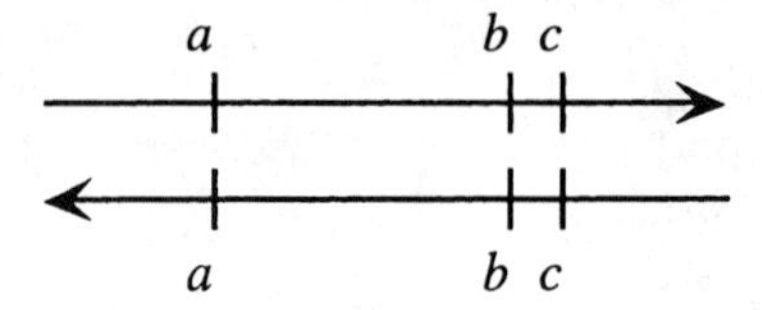

13. The best explanation is that the integrated F factor of the Hfr looped out of the bacterial chromosome abnormally and is now an F′ that contains the pro^+ gene. This F′ is rapidly transferred to F^- cells, converting them to pro^+ (and F^+).

14. The high rate of integration and the preference for the same site originally occupied by the F factor suggest that the F′ contains some homology with the original site. The source of homology could be a fragment of the F factor, or more likely, it is homology with the chromosomal copy of the bacterial gene that is also present on the F′.

15. First carry out a cross between the Hfr and F^-, and then select for colonies that are *ala*$^+$ *str*^{r}. If the Hfr donates the *ala* region late, then redo the cross but now interrupt the mating early and select for *ala*$^+$. This selects for an F′, since this Hfr would not have transferred the *ala* gene early.

If the Hfr instead donates this region early, then use a Rec^- strain that cannot incorporate a fragment of the donor chromosome by recombination. Any *ala*$^+$ colonies from the cross should then be used in a second mating to another *ala*$^-$ strain to see whether they can donate the *ala* gene easily, which would indicate that there is F′ *ala*. (This would also require another marker to differentiate the donor and recipient strains. For example, the *ala*$^-$ strain could be tetracycline^r and selection would be for *ala*$^+$ *tet*^{r}.)

16. a., b.

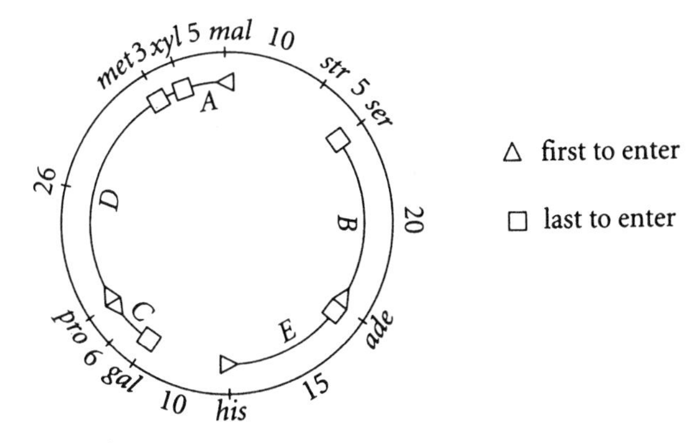

c. A: Select for *mal*$^+$
B: Select for *ade*$^+$
C: Select for *pro*$^+$
D: Select for *pro*$^+$
E: Select for *his*$^+$

17. a. If the two genes are far enough apart to be located on separate DNA fragments, then the frequency of double transformants should be the product of the frequency of the two single transformants, or (4.3%) × (0.40%) = 0.017%. The observed double transformant frequency is 0.17%, a factor of 10 greater than expected. Therefore, the two genes are located close enough together to be cotransformed at a rate of 0.17%.

b. Here, when the two genes must be contained on separate pieces of DNA, the rate of cotransformation is much lower, confirming the conclusion in part a.

18. a. To determine which genes are close, compare the frequency of double transformants. Pairwise testing gives low values whenever B is involved but fairly high rates when any drug but B is involved. This suggests that the gene for B resistance is not close to the other three genes.

b. To determine the relative order of genes for resistance to A, C, and D, compare the frequency of double and triple transformants. The frequency of resistance to AC is approximately the same as resistance to ACD. This strongly suggests that D is in the middle. Also, the frequency of AD co-resistance is higher than AC (suggesting that the gene for A resistance is closer to D than to C), and the frequency of CD is higher than AC (suggesting that C is closer to D than to A).

19. In a small percent of the cases, *gal*$^+$ transductants can arise by recombination between the *gal*$^+$ DNA of the λdgal transducing phage and the *gal*$^-$ gene on the chromosome. This will generate *gal*$^+$ transductants without phage integration.

20. a. This appears to be specialized transduction. It is characterized by the transduction of specific markers based on the position of the integration of the prophage. Only those genes near the integration site are possible candidates for misincorporation into phage particles that then deliver this DNA to recipient bacteria.

b. The only media that supported colony growth were those lacking either cysteine or leucine. These selected for *cys*$^+$ or *leu*$^+$ transductants and indicate that the prophage is located in the *cys-leu* region.

21. a., b.

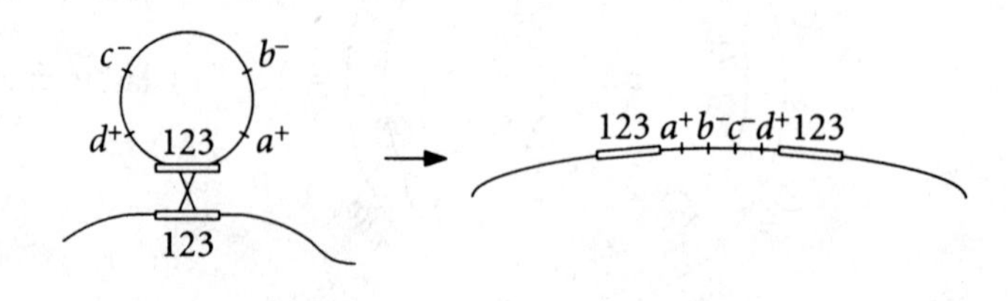

c.

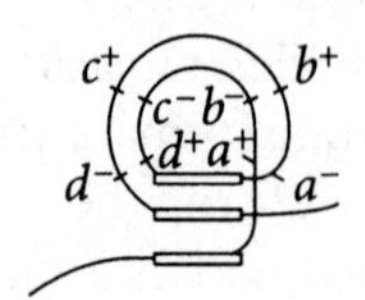

d.

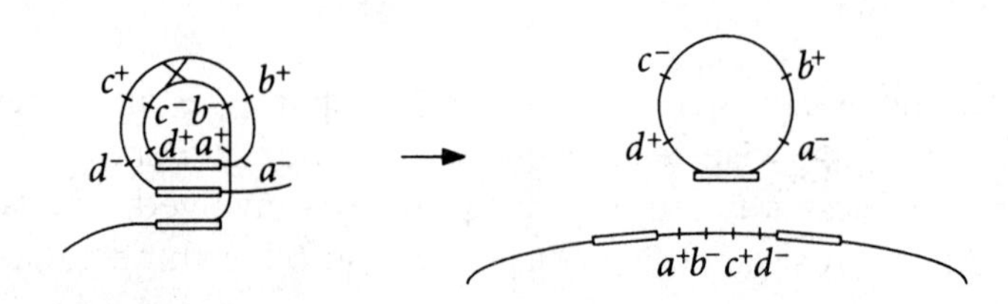

22. If a compound is not added and growth occurs, the *E. coli* recipient cell must have received the wild-type genes for production of those nutrients by transduction. Thus, the BCE culture selects for cells that are now a^+ and d^+, the BCD culture selects for cells that are a^+ and e^+, and the ABD culture selects for cells that are c^+ and e^+. These genes can be aligned, see below, to give the map order of *d a e c*. (Notice that *b* is never cotransduced and is therefore distant from this group of genes.)

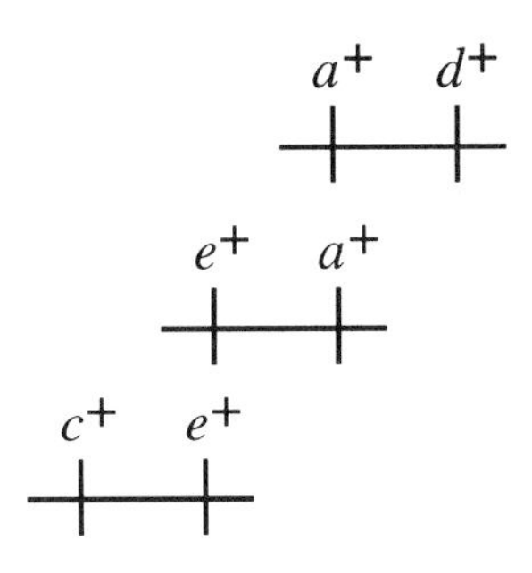

23. a. This is simply calculated as the percentage of pur^+ colonies that are also nad^+:

$$= 100\%(3 + 10)/50 = 26\%$$

b. This is calculated as the percentage of pur^+ colonies that are also pdx^-:

$$= 100\%(10 + 13)/50 = 46\%$$

c. *pdx* is closer, as determined by cotransduction rates.

d. From the cotransduction frequencies, you know that *pdx* is closer to *pur* than *nad* is, so there are two gene orders possible: *pur pdx nad* or *pdx pur nad.* Now, consider how a bacterial chromosome that is $pur^+\ pdx^+\ nad^+$ might be generated given the two gene orders: if *pdx* is in the middle, 4 crossovers are required to get $pur^+\ pdx^+\ nad^+$; if pur is in the middle, only 2 crossovers are required (see below). The results indicate that there are fewer $pur^+\ pdx^+\ nad^+$ transductants than any other class suggesting that this class is "harder" to generate than the others. This implies that *pdx* is in the middle and the gene order is *pur pdx nad.*

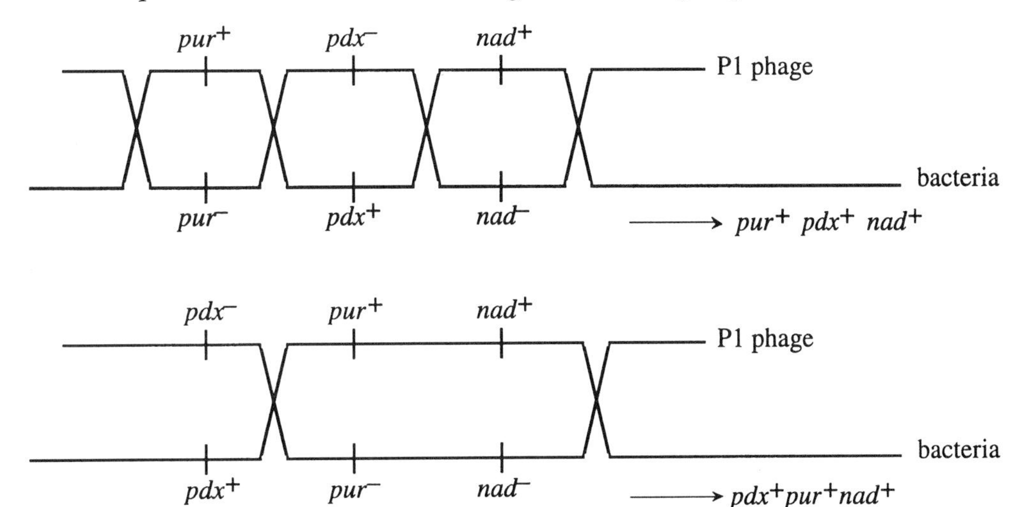

24. a. Owing to the medium used, all colonies are *cys*$^+$ but either + or – for the other two genes.

b. (1) *cys*$^+$ *leu*$^+$ *thr*$^+$ and *cys*$^+$ *leu*$^+$ *thr*$^-$ (supplemented with threonine)
(2) *cys*$^+$ *leu*$^+$ *thr*$^+$ and *cys*$^+$ *leu*$^-$ *thr*$^+$ (supplemented with leucine)
(3) *cys*$^+$ *leu*$^+$ *thr*$^+$ (no supplements)

c. Because none grew on minimal medium, no colony was *leu*$^+$ *thr*$^+$. Therefore, medium (1) had *cys*$^+$ *leu*$^+$ *thr*$^-$, and medium (2) had *cys*$^+$ *leu*$^-$ *thr*$^+$. The remaining cultures were *cys*$^+$ *leu*$^-$ *thr*$^-$, and this genotype occurred in 100% – 56% – 5% = 39% of the colonies.

d. *cys* and *leu* are cotransduced 56% of the time, while *cys* and *thr* are cotransduced only 5% of the time. This indicates that *cys* is closer to *leu* than it is to *thr*. Since no *leu*$^+$ *cys*$^+$ *thr*$^+$ cotransductants are found, it indicates that *cys* is in the middle.

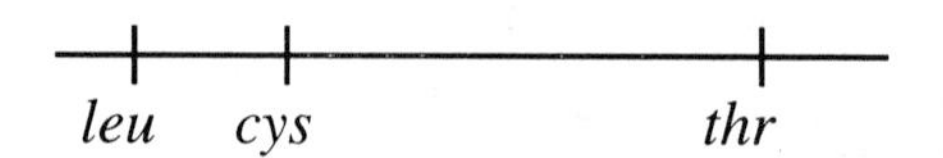

25. a. The parental genotypes are + + + and *m r tu*. For determining the *m–r* distance, the recombinant progeny are

m + *tu*	162
m + +	520
+ *r tu*	474
+ *r* +	172
	1328

Therefore the map distance is 100%(1328)/10,342 = 12.8 m.u.

Using the same approach, the *r–tu* distance is 100%(2152)/10,342 = 20.8 m.u., and the *m–tu* distance is 100%(2812)/10,342 = 27.2 m.u.

b. Since genes *m* and *tu* are the farthest apart, the gene order must be *m r tu*.

c. The coefficient of coincidence (c.o.c.) compares the actual number of double crossovers to the expected number (where c.o.c. = observed double crossovers/expected double crossovers). For these data, the expected number of double recombinants is (0.128)(0.208)(10,342) = 275. Thus, c.o.c. = (162 + 172)/275 = 1.2. This indicates that there are more double crossover events than predicted and suggests that the occurrence of one crossover makes a second crossover between the same DNA molecules more likely to occur.

10 Recombinant DNA Technology

1. ***Unpacking the Problem***

   a. Of the two discussed in the text, pBR322 is the closer.

   b. The single *Hin*dIII site in pBP1 allows for a simple opening up of the plasmid so that a DNA fragment made with *Hin*dIII can be inserted.

   c. It is important because it allows for screening for insertions of DNA into the plasmid. If the plasmid simply recircularizes, the transformed bacteria will be tetracycline-resistant. If the plasmid contains "foreign" DNA, the *tet* gene will be disrupted and the strain will be tetracycline sensitive.

   d. Insertion of donor DNA into the plasmid disrupts the *tet* gene. It is not relevant to the problem but was important in the construction of the library.

   e. A library is a large collection of cloned DNA maintained within easily cultured vectors. For this question, the source of donor DNA was *Hin*dIII-digested fruit fly genomic DNA, and the vector was pBP1. Although it is not relevant to this question, the source of donor DNA is often key to the type of research being conducted.

   f. The gene of interest would have been "found" by using a probe composed of the gene's sequence (typically just a small region is required). This could have been synthesized by "guessing" the DNA sequence on the basis of the gene product's amino acid sequence or by homology to a similar gene from another organism, etc. For this particular question, how this clone was identified does not matter.

g. An electrophoretic gel is an apparatus to separate fragments of DNA by their size. Generally, the mix of DNA fragments is forced to migrate through an agarose gel by an electric field that is negative at the end where the DNA is placed and positive at the far end. Since DNA is negatively charged, it will move to the far end but at rates that are inversely proportional to its size: small fragments will move more rapidly than large.

h. Ethidium bromide binds to DNA and fluoresces when exposed to UV light. It is used to visualize the location of the various DNA fragments within the gel.

i. The DNA from this gel is not "blotted" onto filter paper in this problem. If it had been, it would have been a Southern blot (since DNA was in the gel).

j. In this gel, DNA molecules of different sizes bound to ethidium bromide are visible under UV illumination.

k. There is only one linear fragment generated when a circle is cut once.

l. If cut twice, two linear fragments are generated.

m. There is a one-to-one relationship between the number of sites cut in a circular plasmid and the number of fragments generated.

n. Since the two enzymes will cut the DNA independently, the total number of fragments will be $n + m$.

o. They were loaded into the wells located at the top of the diagram.

p. Smaller fragments move more rapidly and travel farther per unit time than larger fragments.

q. All the control lanes contain 5 kb of DNA, the size of the plasmid. Both *Hin*dIII and *Eco*RV cut the plasmid once but at separate locations, as seen in the lane of the double digest (both single digests generate a single band while the double digest generates two). The lanes with the clone 15-containing plasmid always add up to 7.5 kb, indicating that the donor DNA is 2.5 kb.

r. No. The 5-kb plasmid is cut twice, and the resulting fragments must add up to the total length.

s. No. They represent the cloned DNA cut out from the plasmid by *Hin*dIII and then cut once again by *Eco*RV. The sum of these fragments must equal the whole.

t. It tells you that the fragment that disappeared also contains a restriction site for the second enzyme.

u. A probe will hybridize to any fragments to which it is complementary in sequence.

v. If the two vectors are nonhomologous, the only hybridization observed will be because the gene of interest from the one species is complementary to the gene of interest in the other.

Solution to the Problem

a.

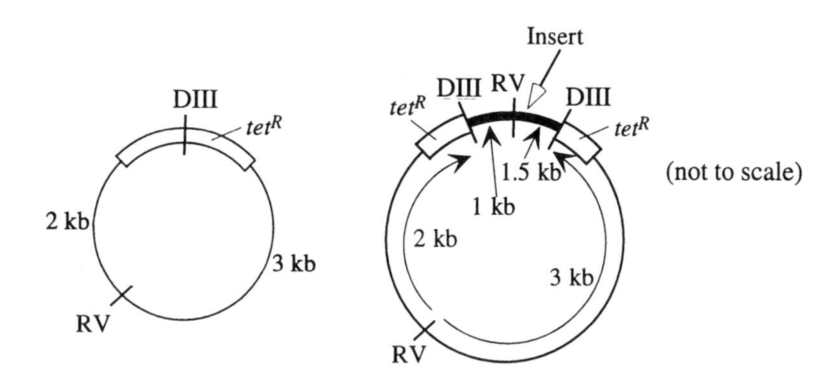

b. The tet^R gene used as a probe will detect only those bands that contain tet^R DNA. Thus, all bands in control lanes will have sequences complementary to the probe. For the clone 15 digests, the *Hin*dIII 5-kb band will be radioactive and both *Eco*RV bands will be radioactive. For the *Hin*dIII + *Eco*RV double digest, the 3-kb and 2-kb bands will be radioactive.

c. The homologous gene used as a probe will detect only those fragments containing the gene of interest. Thus, no bands will be radioactive in the control lanes, and the clone 15 lanes will all have at least one radioactive band. For *Hin*dIII, the 2.5-kb band (the insert) will be radioactive. For *Eco*RV, the 4.5-kb and 3.0-kb bands will be radioactive. For the *Hin*dIII + *Eco*RV double digest, the 1.5-kb and 1-kb bands will be radioactive.

2. This problem assumes a random and equal distribution of nucleotides. Thus, a specific sequence of length n will occur on average once in every 4^n base pairs. GTTAAC occurs, on average, every 4^6 bases, which is 4.096 kb. GGCC occurs, on average, every 4^4 bases, which is 0.256 kb.

3. The data indicate that *Eco*RI fragments 1 and 4 contain no *Hin*dII sites, fragment 3 contains one *Hin*dII site, and fragment 2 contains two sites. Conversely, *Hin*dII fragments A, B, and D all contain one *Eco*RI site, and fragment C contains none. Fragment D contains fragments 1 and 3_1; fragment A contains fragments 3_2 and 2_1; fragment C is the same as fragment 2_2; and fragment B contains fragments 2_3 and 4. The only map consistent with these data is

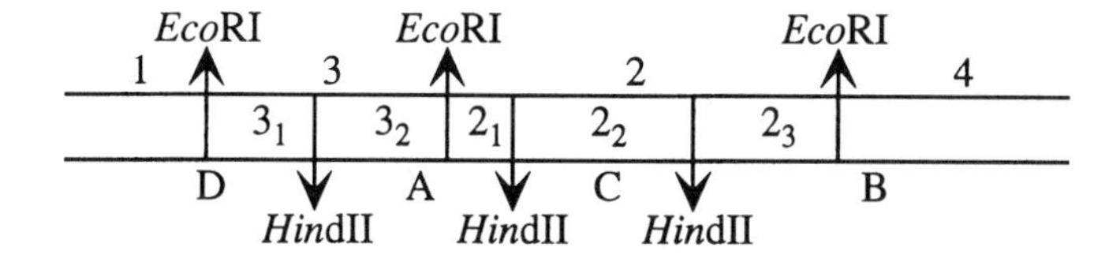

4. **a.** Since the actin protein sequence is known, a probe could be synthesized by "guessing" the DNA sequence based on the amino acid sequence. (This works best if there is a region of amino acids that can be coded with minimal redundancy.) Alternatively, the gene for actin cloned in another

species can be used as a probe to find the homologous gene in *Drosophila.* If an expression vector was used, it might also be possible to detect a clone coding for actin by screening with actin antibodies.

b. Hybridization using the specific tRNA as a probe could identify a clone coding for itself.

5. To answer this problem, you must realize what is being visualized. The 8.5 *Eco*RI fragment is radioactive only at one 5′ end, and only fragments containing that end will be seen by autoradiography. When this fragment is cleaved by other restriction enzymes, the longest fragments will have been cut at sites farthest from the radioactive end. In the following figure, if cut at position labeled 2, the fragment will be longer than any fragment cut at 3, 4, or 5 and shorter than any cut at position 1.

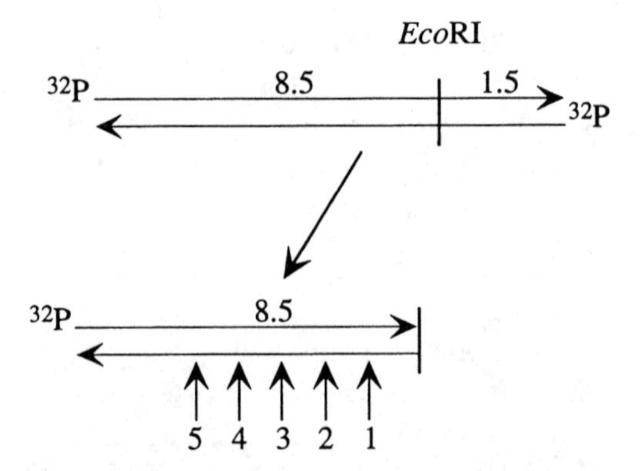

Using this logic, a relative map of the restriction sites of *Hin*dII and *Hae*III for this fragment can be generated.

Reading the gels from the top (and from the farthest to the nearest to the labeled end) the order is (*Eco*RI) - *Hin*dII - *Hae*III - *Hae*III - *Hin*dII - *Hae*III - *Hae*III - *Hae*III - *Hin*dII - *Hin*dII - *Hae*III - *Hin*dII - *Hae*III - labeled end

6. This problem assumes a random and equal distribution of nucleotides.

*Alu*I $(^1/_4)^4$ = on average, once in every 256 nucleotide pairs

*Eco*RI $(^1/_4)^6$ = on average, once in every 4096 nucleotide pairs

*Acy*I $(^1/_4)^4(^1/_2)^2$ = on average, once in every 1024 nucleotide pairs

7. **a.** The double digest indicates that the 5.0-kb *Hin*dIII fragment also contains a *Sma*I site and the 5.5 *Sma*I fragment also contains a *Hin*dIII site. This suggests the following map:

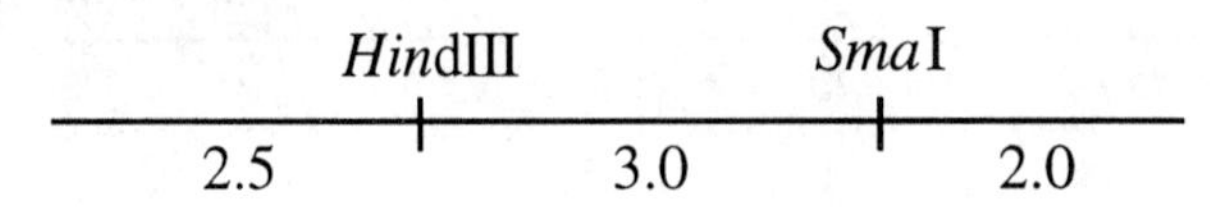

b. Since the only band to disappear is the 3.0-kb fragment, it is the only one that also contains an *Eco*RI site. The appearance of a new 1.5-kb fragment suggests the following:

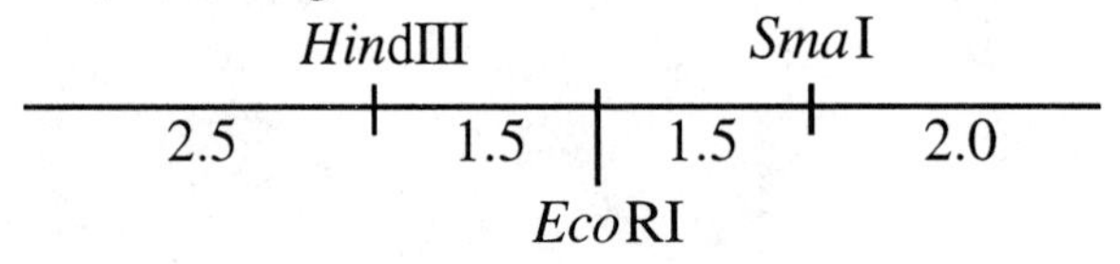

8. **a.** Since the protein is present only after the mRNA has been processed, the mature mRNA must be 1200 nucleotides.

b. The autoradiogram will show only the radioactive RNA molecules. At 2 hours the cDNA does not protect the entire RNA from RNase, but at 10 hours it does. Since the cDNA would not contain sequences complementary to introns, there must be an intron present in the pre-mRNA at 2 hours that is spliced out by 10 hours.

At 2 hours the viral transcript contains an intron sequence that does not hybridize to the cDNA. The RNase removes the single-stranded sequence, leaving behind 500 and 700-base fragments:

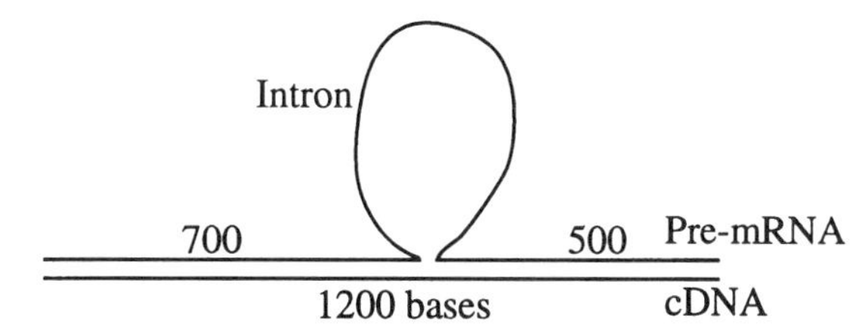

By 10 hours, the intron has been spliced out, and a perfect hybrid forms between the 1200-base viral mRNA and the cDNA:

1200 bases mRNA

1200 bases cDNA

c. It takes a minimum of 2 hours to transcribe and splice the mRNA and then translate it into protein.

9. Create a library of *Podospora* DNA. This would be accomplished by isolating DNA from *Podospora*, cutting the DNA with *Hae*II, mixing with the vector pBR also cut with *Hae*II, ligating the mixture, and transforming *E. coli*. Only those bacteria that contain the plasmid will be tet^R. Of these, those that are kan^S contain plasmids with inserts.

Assuming that the same genes from different species have approximately the same base sequence, the ß-tubulin gene cloned from *Neurospora* can be used as a probe to isolate the ß-tubulin gene from *Podospora*. Identify which clone or clones in the library contain the desired sequence by colony hybridization using the cloned *Neurospora* actin gene as a probe.

10. **a.** There is one *Bgl*II site, and the plasmid is 14 kb.

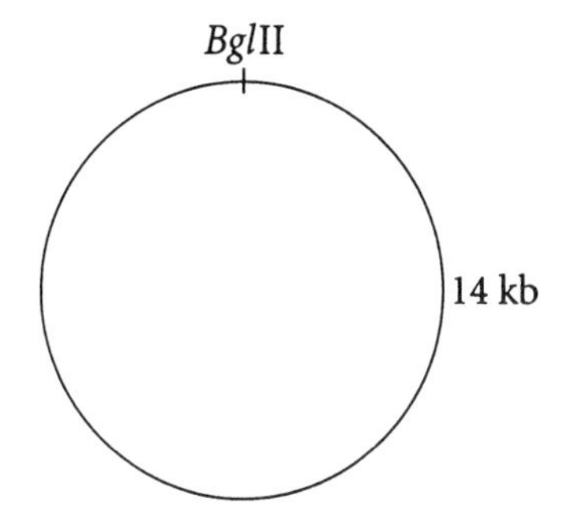

b. There are two *Eco*RV sites.

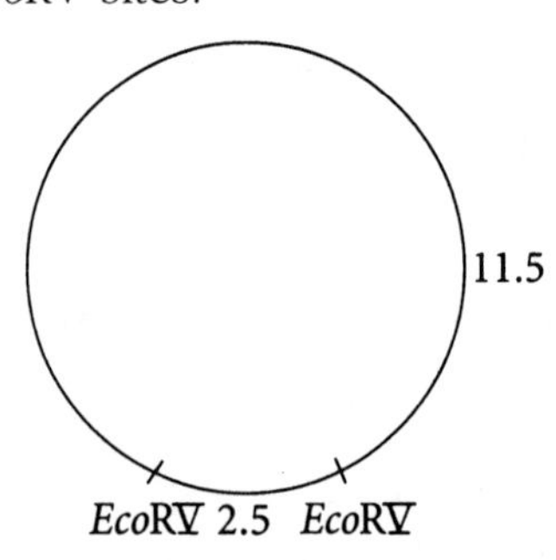

c. The 11.5 kb RV fragment is cut by *Bgl*II. The arrangement of the sites must be as indicated below.

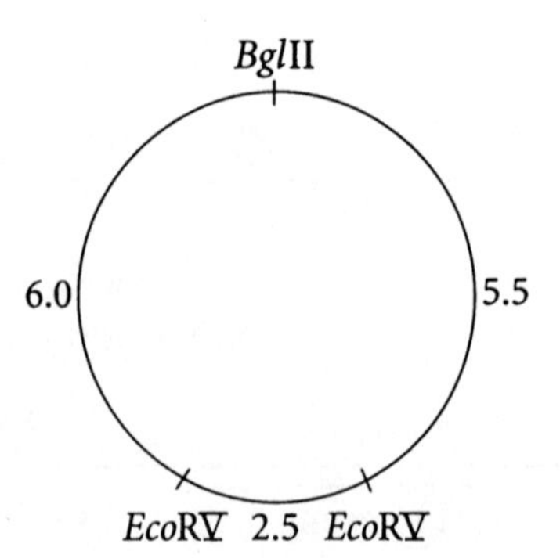

d. The *Bgl*II site must be within the *tet* gene.

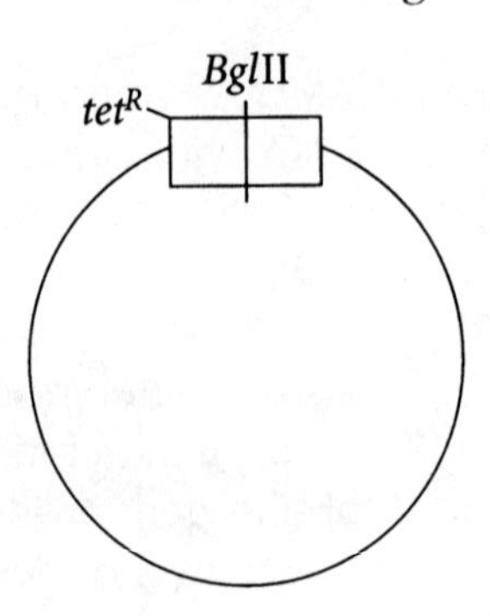

e. There was an insert of 4 kb.

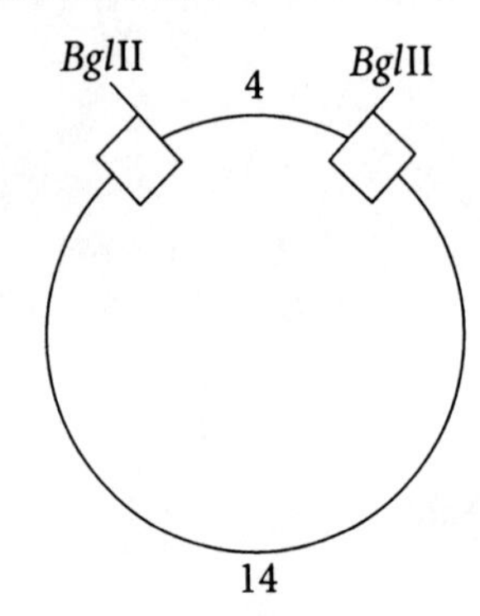

f. There was an *Eco*RV site within the insert.

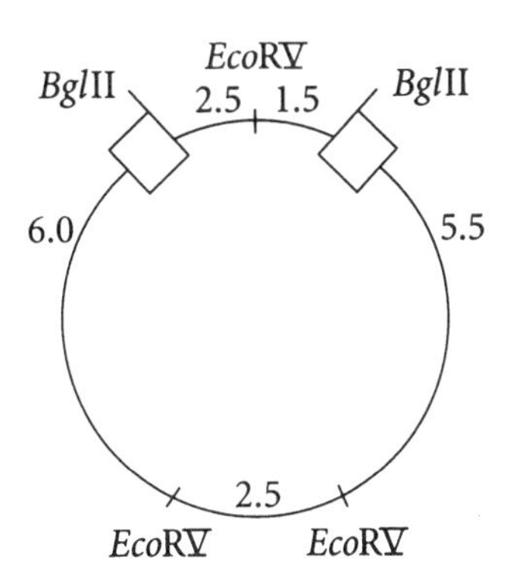

11. a. The restriction map of pBR322 with the mouse fragment inserted is shown below. The 2.5-kb and 3.5-kb fragments would hybridize to the pBR322 probe.

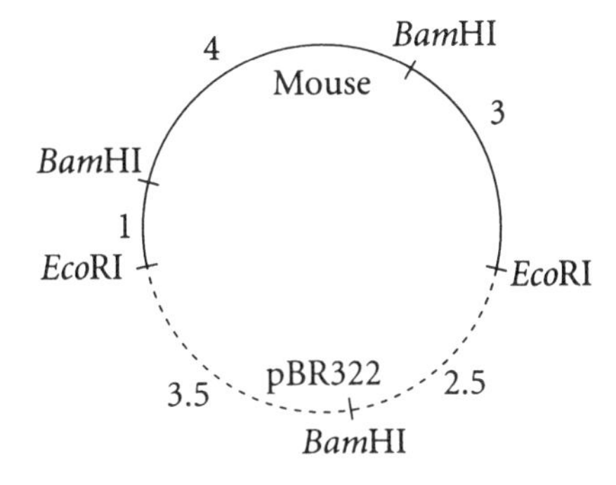

b. A protein 400 amino acids long requires a minimum of 1200 nucleotide bases. Only fragment 3 is long enough (3000 bp) to contain two or more copies of the gene. However, nothing can be said about their orientation.

12. a. To ensure that a colony is not, in fact, a prototrophic contaminant, the prototrophic line should be sensitive to a drug to which the recipient is resistant. A simple additional marker would also achieve the same end.

b. Use a nonrevertible auxotroph as the recipient (such as one containing a deletion).

13. a. The transformed phenotype would map to the same locus. If gene replacement occurred by a double crossing-over event, the transformed cells would not contain vector DNA. If a single crossing-over took place, the entire vector would now be part of the linear *Neurospora* chromosome.

b. The transformed phenotype would map to a different locus than that of the auxotroph if the transforming gene was inserted ectopically (i.e., at another location).

14. Size, translocations between known chromosomes, and hybridization to probes of known location can all be useful in identifying which band on a PFGE gel corresponds to a particular chromosome.

15. a., b.

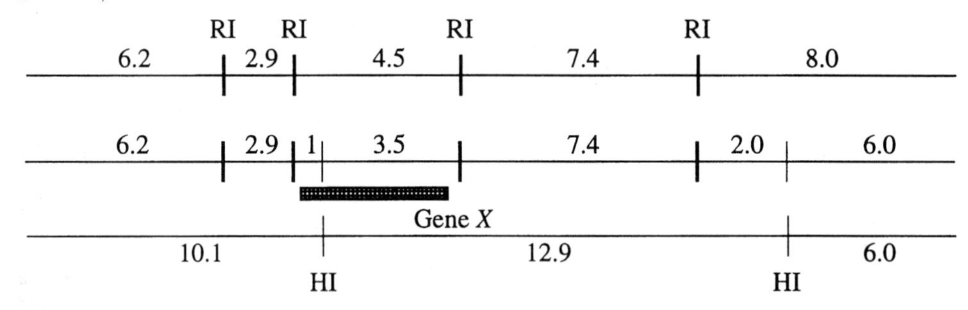

16. a. The total weight of the bands is 2.3 kilobases. The combined digest results in four bands, and *Taq*I results in two bands. This suggests that *Taq*I cuts once in the linearized fragment and that *Hae*III cuts once within each *Taq*I fragment. Note that 0.8 + 0.4 = 1.2, and that 0.6 + 0.5 = 1.2. The map of all the fragments is

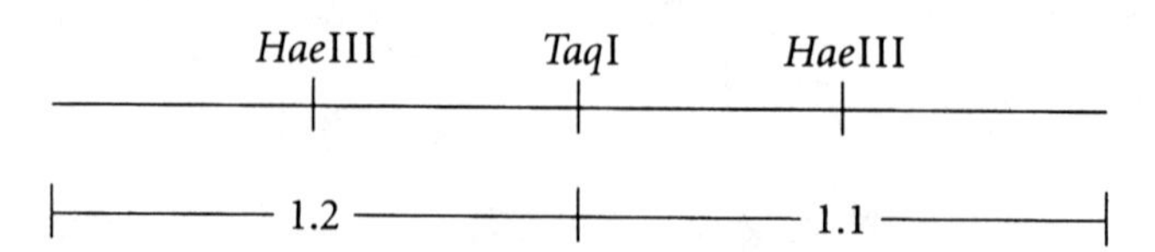

Only the 0.8 and 0.5 fragments contain the ends. Therefore, the complete map is as follows

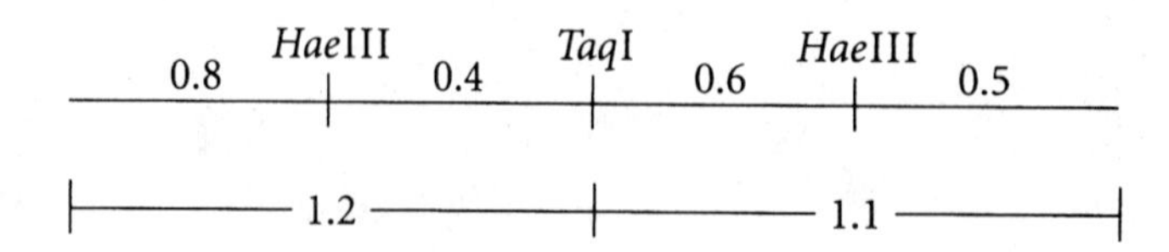

b. There are three bands, only two of which have labeled ends. The map is

*Eco*RI *Bam*HI *Bam*HI *Eco*RI

3.4 3.0 2.2

c. The 1.1 *Taq*I fragment should hybridize to the 2.2 fragment. However, without knowing the number of introns, it is possible that this *Taq*I fragment also contains sequences complementary to the 3.4 genomic fragment.

d. The 3.0 genomic fragment is very likely an intron. That is why the cDNA had no complementary sequences to this fragment.

17. a. If an individual is homozygous for an allele, there should be only one band on all the blots. While this does not prove homozygosity, it is consistent with that finding. On the other hand, two bands on any of these gels does show heterozygosity. Therefore, it is consistent that individuals 1, 3, 4, and 5 are homozygous.

b. Individual 3 makes no RNA. Therefore, the most likely conclusion is that the subject is homozygous for a mutation that blocks transcription.

Individual 4 makes a protein smaller than seen in unaffected people, but the mRNA is of the same size. This suggests that a chain termination (nonsense) mutation occurred.

Individual 5 makes a normal-length transcript but a slightly larger protein. One explanation is that a frameshift mutation occurred that eliminated the normal stop codon. Another explanation is a point mutation that altered the normal stop codon so that it coded for an amino acid.

18. a. Note that the genomic *Hin*dIII fragment is end labeled before the *Bam* digest. This means that the 10 kb *Bam-Hin*dIII fragment is labeled only on its *Hin*dIII end. Only fragments extending from this end toward the *Bam* site are detectable in the partial digest with *Hpa*II (see Problem 5). The genomic map is

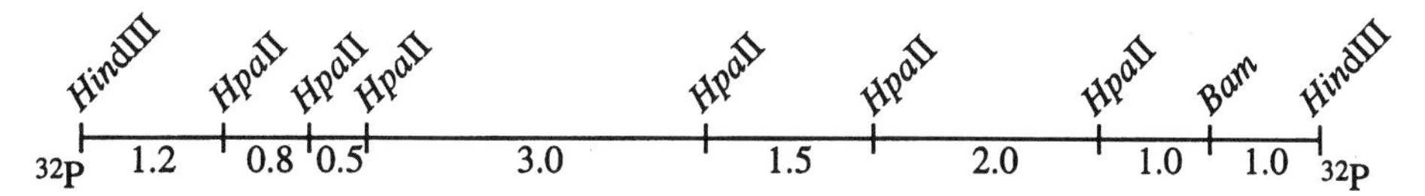

b., c. In the following restriction map of the genomic DNA, the cDNA fragments are marked underneath by boxes and the "gaps" represent introns. There is a single *Hpa*II site within the cDNA, which is also marked. However, the exact ends and sizes of the cDNA fragments are not established. Fragment 2 contains sequences complementary to a 1-kb fragment, but there are two alternatives possible.

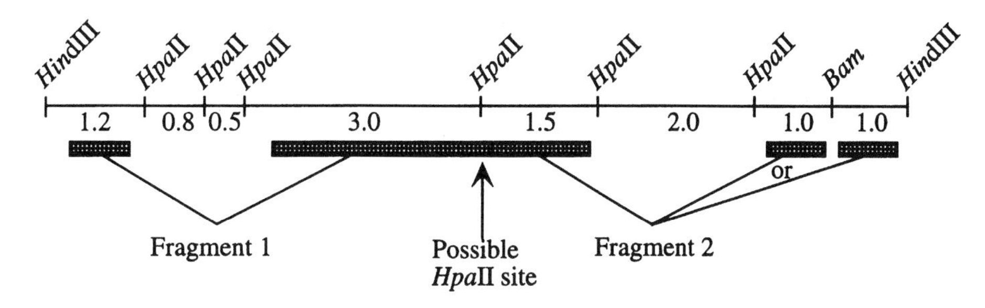

The genomic and cDNA maps differ dramatically because at least two introns have been removed from the genomic DNA to produce the cDNA.

19. a.

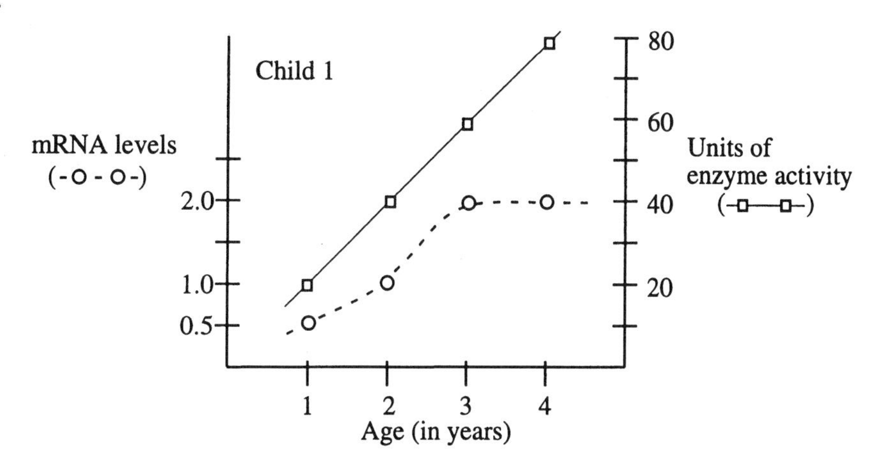

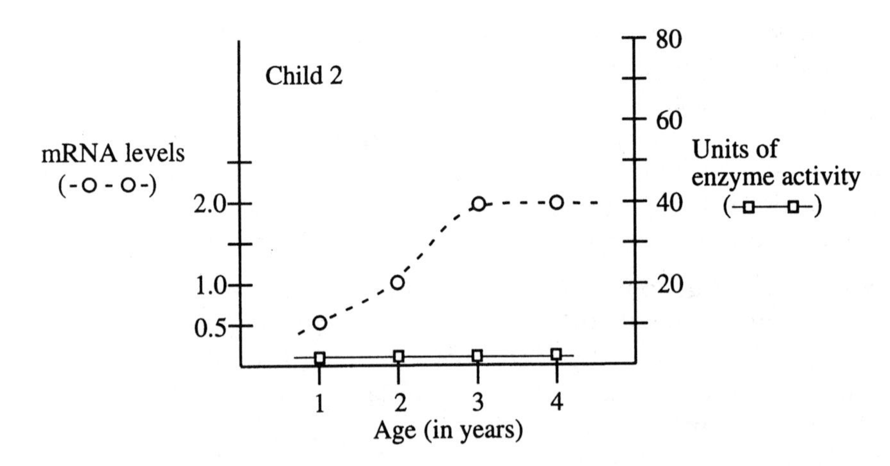

b., **c.** Very low levels of active enzyme are caused by the introduction of an *Xho* site within the D gene. Because the size of the nonfunctional enzyme has not changed, the most likely event was a point mutation within the coding region of the gene that also created a new *Xho* site. It is likely that this point mutation also altered the active site destroying enzyme activity.

d. Individual 1 would be defined as homozygous normal, while individual 2 is homozygous mutant. If either were heterozygous, there would be three bands hybridizing to the probe on the Southern blot.

20. **a.** The gel can be read from the bottom to the top in a 5´-to-3´ direction. The sequence is

5´ TTCGAAAGGTGACCCCTGGACCTTTAGA 3´

b. By complementarity, the template was

3´ AAGCTTTCCACTGGGGACCTGGAAATCT 5´

c. The double helix is

5´ TTCGAAAGGTGACCCCTGGACCTTTAGA 3´
3´ AAGCTTTCCACTGGGGACCTGGAAATCT 5´

d. Open reading frames have no stop codons. There are three frames for each strand, for a total of six possible reading frames. For the strand read from the gel, the transcript would be

5´ UC**UAA**AGGUCCAGGGGUCACCUUUCGAA 3´

And for the template strand

5´ UUCGAAAGG**UGA**CCCCUGGACCUU**UAG**A 3´

Stop codons are in bold and underlined. There are a total of four open reading frames of the six possible.

21. The region of DNA that encodes tyrosinase in "normal" mouse genomic DNA contains two *Eco*RI sites. Thus, after *Eco*RI digestion, three different-sized fragments hybridize to the cDNA clone. When genomic DNA from certain albino mice is subjected to similar analysis, there are no DNA fragments that contain complementary sequences to the same cDNA. This indicates that these mice lack the ability to produce tyrosinase because the DNA that encodes the enzyme must be deleted.

22. Conservatively, the amount of DNA necessary to encode this protein of 445 amino acids is $445 \times 3 = 1335$ base pairs. When compared with the actual amount of DNA used, 60 kb, the gene appears to be roughly 45 times larger than necessary. This "extra" DNA mostly represents the introns that must be correctly spliced out of the primary transcript during RNA processing for correct translation. (There are also comparatively very small amounts of both 5′ and 3′ untranslated regions of the final mRNA that are necessary for correct translation encoded by this 60-kb of DNA.)

11 Applications of Recombinant DNA Technology

1. Plant 1 shows the typical inheritance for a dominant gene that is heterozygous. Assuming kanamycin resistance is dominant to kanamycin sensitivity, the cross can be outlined as follows:

$$kan^R/kan^S \times kan^S/kan^S$$
$$\downarrow$$
$$\tfrac{1}{2}\ kan^R/kan^S$$
$$\tfrac{1}{2}\ kan^S/kan^S$$

This would suggest that the gene of interest would be inserted once into the genome.

Plant 2 shows a 3:1 ratio in the progeny of the backcross. This suggests that there have been two unlinked insertions of the *kan*^{R} gene and presumably the gene of interest as well.

$$kan^{R1}/kan^{S1};\ kan^{R2}/kan^{S2} \times kan^{S1}/kan^{S1};\ kan^{S2}/kan^{S2}$$
$$\downarrow$$
$$\tfrac{1}{4}\ kan^{R1}/kan^{S1};\ kan^{R2}/kan^{S2}$$
$$\tfrac{1}{4}\ kan^{R1}/kan^{S1};\ kan^{S2}kan^{S2}$$
$$\tfrac{1}{4}\ kan^{S1}/kan^{S1};\ kan^{R2}/kan^{S2}$$
$$\tfrac{1}{4}\ kan^{S1}/kan^{S1};\ kan^{S2}/kan^{S2}$$

2. PFGE separates large DNA molecules (small chromosomes) by nature of their size. Unless the overall size of the DNA molecule (chromosome) has been changed, it will migrate in the same relative position.

a. The size of chromosome 1 has not been changed, so there will still be the same 7 bands detected, one for each chromosome.

b. The same as (a).

c. Both chromosomes 1 and 7 have been altered. Compared with wild type, the largest and smallest chromosomes will disappear and two new intermediate bands will appear (unless they happen to co-migrate with any of the other wild-type bands).

d. One band will be larger than expected and one will be smaller when compared with wild type.

e. The same as wild type, 7 bands. The "extra" chromosome will co-migrate with its homolog.

f. It is the number of different chromosomes, not the ploidy, that is observed on the gel. This will be the same as wild type (although *Neurospora* is typically haploid).

g. Compared with wild type, the largest band would disappear and be replaced by an even larger band.

3. One way to approach this question is to use the cloned DNA as a probe to see if the gene is transcribed in the nonphotosynthetic tissues. mRNA from various tissues would be isolated and separated on a gel, and a Northern blot performed with the clone as the probe. A band would be detected in only those tissues where the gene is transcribed.

4. There are four patterns of RFLPs possible: the strain 1 pattern; the strain 2 pattern; and the two recombinant patterns of strain 1 for probe A, strain 2 for probe B and strain 1 for probe B, strain 2 for probe A. When probed with both A and B, the following is expected:

The pattern of bands tells you the "genotype" of each offspring. Although you are told that both RFLPs are on chromosome 5, you do not know if they

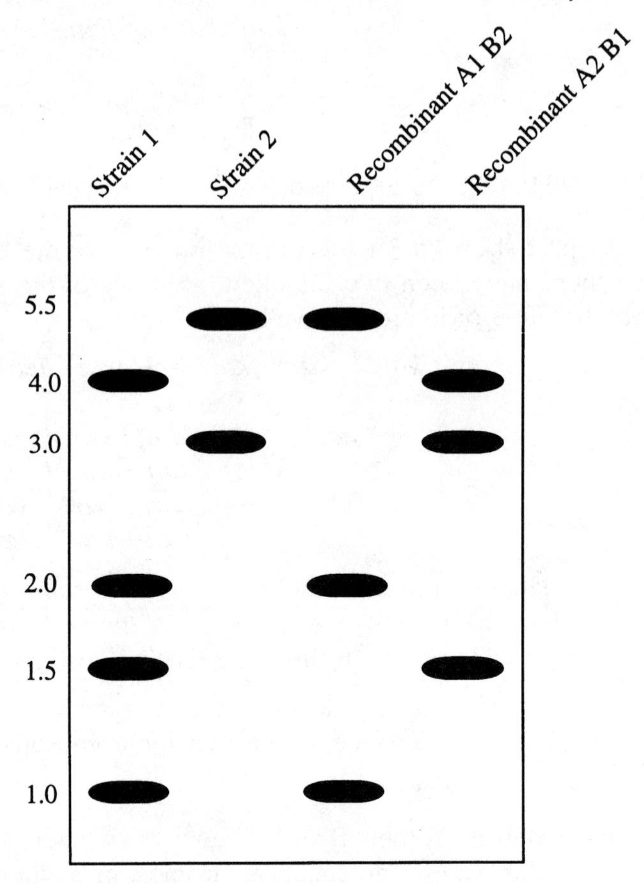

are on the same or opposite sides of the centromere. The patterns of M_{II} segregation would help distinguish between these two possibilities. For illustration, assume that the two RFLPs are on opposite sides of and each 10 m.u. away from the centromere. For this case, 20% of the asci will show M_{II} segregation for the A RFLP and 20% will show M_{II} segregation for the B RFLP. To analyze this, each spore would have to be grown separately, and from each colony, DNA would have to be isolated, digested with *Pst*I, and submitted to Southern blot testing. The patterns of bands observed would determine which RFLP marker each progeny inherited. Just considering the RFLP identified by clone A, the following asci would be observed:

A1	A2	A1	A1	A2	A2
A1	A2	A1	A1	A2	A2
A1	A2	A2	A2	A1	A1
A1	A2	A2	A2	A1	A1
A2	A1	A1	A2	A2	A1
A2	A1	A1	A2	A2	A1
A2	A1	A2	A1	A1	A2
A2	A1	A2	A1	A1	A2
40%	40%	5%	5%	5%	5%

The distance of the marker to the centromere would be calculated as $^1/_2$ of those asci showing M_{II} segregation. The same results would be observed for the B RFLP.

The map distance between the two RFLPs could be calculated using the formula

$$RF = {}^1/_2T + NPD$$

Tetratypes (T) will look like

A1B1	A1B2	A1B1	A1B2	A1B1	A1B1	A2B1	A2B1
A1B1	A1B2	A1B1	A1B2	A1B1	A1B1	A2B1	A2B1
A1B2	A1B1	A1B2	A1B1	A2B1	A2B1	A1B1	A1B1
A1B2	A1B1	A1B2	A1B1	A2B1	A2B1	A1B1	A1B1
A2B1	A2B2	A2B2	A2B1	A2B2	A1B2	A2B2	A1B2
A2B1	A2B2	A2B2	A2B1	A2B2	A1B2	A2B2	A1B2
A2B2	A2B1	A2B1	A2B2	A1B2	A2B2	A1B2	A2B2
A2B2	A2B1	A2B1	A2B2	A1B2	A2B2	A1B2	A2B2

and their "upside-down" versions, as well.

And nonparental ditypes (NPD) will look like

A1B2	A1B2	A1B2
A1B2	A1B2	A1B2
A1B2	A2B1	A2B1
A1B2	A2B1	A2B1
A2B1	A1B2	A2B1
A2B1	A1B2	A2B1
A2B1	A2B1	A1B2
A2B1	A2B1	A1B2

and their "upside-down" versions, as well.

5. Assuming that the DNA from this region is cloned, it could be used as a probe to detect this RFLP on Southern blots. DNA from individuals within this pedigree would be isolated (typically from blood samples containing white blood cells) and restricted with *Eco*RI, and Southern blots would be performed. Individuals with this mutant CF allele would have one band that would be larger (owing to the missing *Eco*RI site) when compared with wild type. Individuals that inherited this larger *Eco*RI fragment would, at minimum, be carriers for cystic fibrosis. In the specific case discussed in this problem, a woman that is heterozygous for this specific allele marries a man that is heterozygous for a different mutated CF allele. Just knowing that both are heterozygous, it is possible to predict that there is a 25% chance of their child's having CF. However, since the mother's allele is detectable on a Southern blot, it would be possible to test whether the fetus inherited this allele. DNA from the fetus (through either CVS or amniocentesis) could be isolated and tested for this specific *Eco*RI fragment. If the fetus did not inherit this allele, there would be a 0% chance of its having CF. On the other hand, if the fetus inherited this allele, there would be a 50% chance the child will have CF.

6. The typical procedure is to "knock out" the gene in question and then see if there is any observable phenotype. One methodology to do this is described in Figure 11.4 in the companion text. A recombinant vector carrying a selectable gene within the gene of interest is used to transform yeast cells. Grown under appropriate conditions, yeast that have incorporated the marker gene will be selected. Many of these will have the gene of interest disrupted by the selectable gene. The phenotype of these cells would then be assessed to determine gene function.

7. PFGE separates chromosomes by size. After electrophoresis, Southern blot the gel and probe with radioactive copies of the cloned gene. The clones will hybridize to the band that corresponds to the chromosome they were originally from.

8. The promoter and control regions of the plant gene of interest must be cloned and joined in the correct orientation with the glucuronidase gene. This places the reporter gene under the same transcriptional control as the gene of interest. Figure 11-15 in the companion text discusses the methodology used to create transgenic plants. Transform plant cells with the reporter gene construct, and as discussed in the figure, grow into transgenic plants. The glucuronidase gene will now be expressed in the same developmental pattern as the gene of interest and its expression can easily be monitored by bathing the plant in an X-Gluc solution and assaying for the blue reaction product.

9. **a.** The cross is B/b; $RFLP^1/RFLP^2 \times b/b$; $RFLP^2/RFLP^2$, where B = bent tail, b = wild type, $RFLP^1$ = 1.7-kb *Hin*dIII fragment, and $RFLP^2$ = 3.8-kb *Hin*dIII fragment.

The bent-tail progeny from this cross are

40% B/b $RFLP^2/RFLP^2$
60% B/b $RFLP^1/RFLP^2$

If the RFLP is unlinked to the bent-tail locus, the two markers should segregate independently. This is not what is observed. $RFLP^1$ appears to be linked to the *B* allele. If enough progeny were observed to have confidence in the observed ratios, the data suggest that the two markers are 40 m.u. apart (there is 40% recombination).

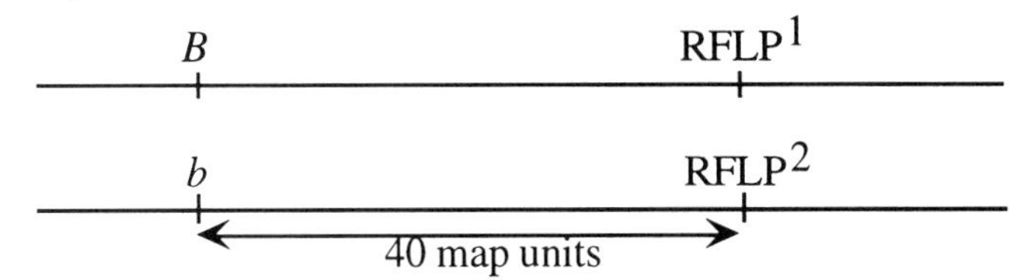

b. The wild-type progeny should be

60% "parental"	$b\,RFLP^2/b\,RFLP^2$
40% "recombinant"	$b\,RFLP^1/b\,RFLP^2$

10. a. The two strains of yeast were crossed (*A1 B1* × *A2 B2*) and the meiotic products were analyzed. If the two RFLP markers are on separate chromosomes, they should assort independently. The actual data are

Spore type	RFLPs	Frequency
1	*A1 B2*	15%
2	*A2 B1*	15%
3	*A1 B1*	35%
4	*A2 B2*	35%

The markers are not assorting randomly. Spore types 1 and 2 are recombinants and types 3 and 4 are parentals. The RFLPs are 30 m.u. apart.

b.

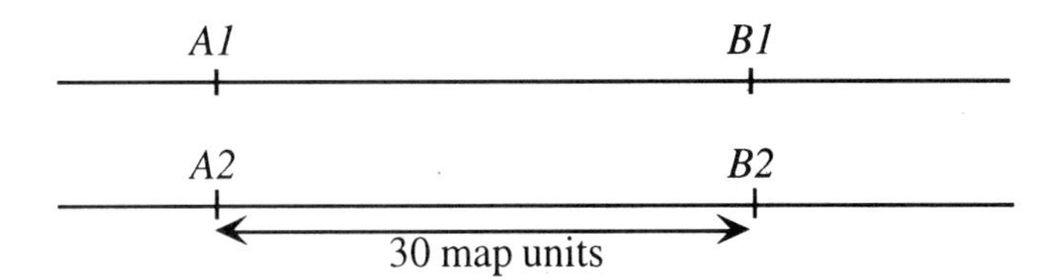

11. a., b. During Ti plasmid transformation, the kanamycin gene will insert randomly into the plant chromosomes. Colony A, when selfed, has $^3/_4$ kanamycin-resistant progeny, and colony B, when selfed, has $^{15}/_{16}$ kanamycin-resistant progeny. This suggests that there was a single insertion into one chromosome in colony A and two independent insertions on separate chromosomes in colony B. This can be schematically represented by showing a single insertion within one of the pair of chromosome "A" for colony A

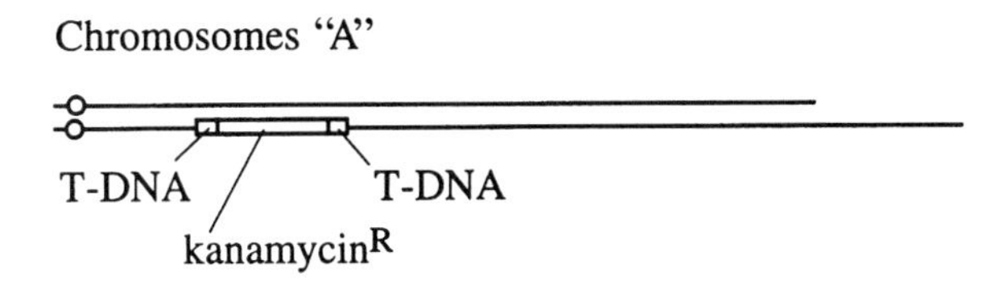

and two independent insertions into one of each of the pairs of chromosomes "B" and "C" for colony B

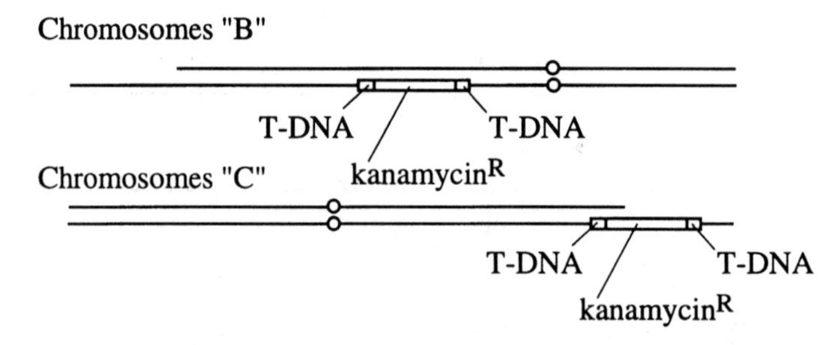

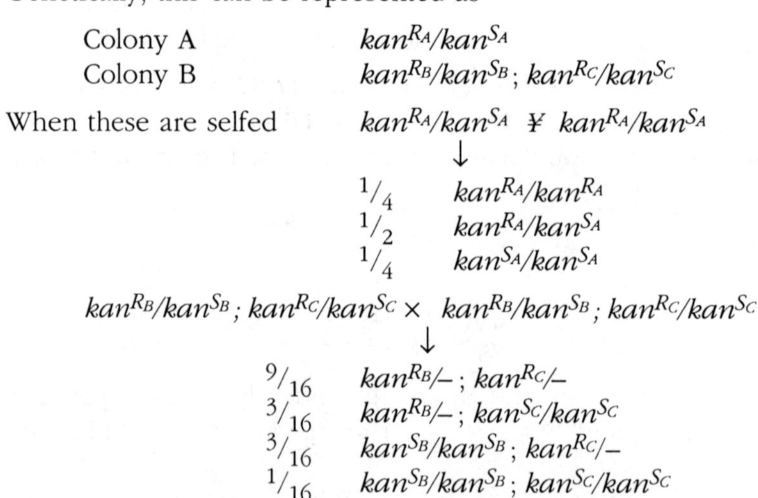

Genetically, this can be represented as

Colony A	kan^{R_A}/kan^{S_A}
Colony B	kan^{R_B}/kan^{S_B}; kan^{R_C}/kan^{S_C}
When these are selfed	kan^{R_A}/kan^{S_A} ¥ kan^{R_A}/kan^{S_A}

↓

$^1/_4$ kan^{R_A}/kan^{R_A}
$^1/_2$ kan^{R_A}/kan^{S_A}
$^1/_4$ kan^{S_A}/kan^{S_A}

kan^{R_B}/kan^{S_B}; kan^{R_C}/kan^{S_C} × kan^{R_B}/kan^{S_B}; kan^{R_C}/kan^{S_C}

↓

$^9/_{16}$ $kan^{R_B}/–$; $kan^{R_C}/–$
$^3/_{16}$ $kan^{R_B}/–$; kan^{S_C}/kan^{S_C}
$^3/_{16}$ kan^{S_B}/kan^{S_B}; $kan^{R_C}/–$
$^1/_{16}$ kan^{S_B}/kan^{S_B}; kan^{S_C}/kan^{S_C}

12. a. Yeast plasmids can exist free in the cytoplasm or can integrate into a chromosome; however, the patterns of inheritance will differ for the two states. Free plasmids are present in multiple copies and will be distributed to all progeny. This is observed in crosses with YP1 transformed cells: YP1 *leu*$^+$ × *leu*$^-$ all progeny are *leu*$^+$ and all have vector DNA. Crosses with YP2 transformed cells show simple Mendelian inheritance and suggest that this plasmid has integrated into the yeast chromosome.

b. For YP1 transformed cells, the circular (and free) plasmid will be linearized by the single restriction cut, and when probed, a single band will be present on the Southern blot. Since the YP2 plasmid is integrated, a single cut within the plasmid will generate two fragments that contain plasmid DNA. However, the size of these fragments will depend on where in the genome other sites exist for this same restriction enzyme. This is schematically shown on page 115:

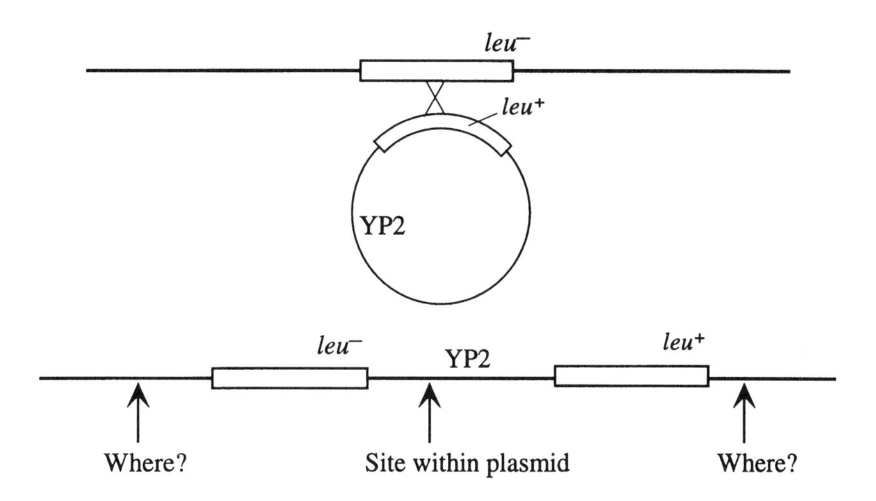

13. a. If the plasmid never integrates, the linear plasmid will be cut once by *Xba*I and two fragments will be generated that will both hybridize to the *Bgl*II probe. The autoradiogram will show two bands whose combined length will equal the full length of the plasmid.

b. If the plasmid integrates occasionally, most cells will still have free plasmids and these will be indicated by the two bands mentioned above. However, when the plasmid is integrated, two bands will still be generated, but their sizes will vary based on where other genomic *Xba*I sites are relative to the insertion point (see the figure for Problem 12b for similar logic). If integration is random, many other bands will be observed, but if it's at a specific site, only two other bands will be detected.

14. Assume the grandfather (I-1) is heterozygous for the dominant Huntington disease allele (which is normally much more likely for this rare disease). Assigning genotypes where possible, the following can be stated

I-1 $H\,\text{RFLP}^{\text{A}}\,/\,h\,\text{RFLP}^{\text{A}}$ × $h\,\text{RFLP}^{\text{B}}/h\,\text{RFLP}^{\text{B}}$ I-2

where H = Huntington disease allele, h = wild-type allele, RFLP^{A} = 3.7- and 1.2-kb fragments, and RFLP^{B} = 4.9-kb fragment.

All other members of this pedigree can also be assigned:

II-1 and II-4	? $\text{RFLP}^{\text{A}}/h\,\text{RFLP}^{\text{B}}$
II-2	$h\,\text{RFLP}^{\text{B}}/h\,\text{RFLP}^{\text{B}}$
II-3	$h\,\text{RFLP}^{\text{A}}/h\,\text{RFLP}^{\text{A}}$

where the ? indicates H or h, both being equally likely. It cannot be determined which chromosome II-1 and II-4 inherited from their father (I-1). Both have a 50% chance of being at risk for Huntington disease (i.e., inheriting the $H\,\text{RFLP}^{\text{A}}$ chromosome).

For fetus III-1, RFLP^{A} was inherited from its father because RFLP^{B} must have been inherited from its mother.

For fetus III-2, $RFLP^B$ was inherited from its mother because $RFLP^A$ must have been inherited from its father.

Since the RFLP being analyzed in this question is 4 m.u. distant from the gene, recombination between the two must be considered. For fetus III-1, it is the probability of inheriting the *H* allele from its father that is relevant and for fetus III-2 it is the probability of inheriting the *H* allele from its mother that is relevant. Both of these parents have a $1/2$ chance of having inherited this allele, and in both cases, it would be linked to $RFLP^A$. Since fetus III-1 has inherited $RFLP^A$, there is $1/2 \times 96\%$ (the chance that $RFLP^A$ is still linked to *H*) = 48% chance that it has inherited Huntington disease. Fetus III-2 has inherited the $RFLP^B$ allele from its mother so there is $1/2 \times 4\%$ (the chance that $RFLP^B$ is now linked to *H*) = 2% chance that it has inherited Huntington disease.

15. *Unpacking the Problem*

a. Hyphae in *Neurospora* are the threads of cells that grow out from the original ascospore. Therefore, **hyphal extension** refers to the pattern of these threads and the distance that they grow. Since this process is due to cell growth (and its control) and cell shape, anyone interested in a vast array of cell biological issues might be interested in the genes identified by such screens.

b. **Mutational dissection** is the attempt to identify all the genes and gene products involved in a particular process. In this experiment, the goal is to identify all the genes that can mutate to a small-colony phenotype by random insertion of unrelated DNA (in this case a bacterial plasmid with a selectable marker).

c. *Neurospora* is a haploid organism, and this is relevant to this problem. What might otherwise be a recessive mutation in a diploid organism (typical of gene knockouts) would instead be immediately expressed in a haploid one.

d. The source of the DNA is not relevant to the problem so long as it contains a selectable marker for the organism being transformed. The ease of growing, manipulating, and isolating bacterial plasmids makes them an attractive choice.

e. Transformation, as originally discovered by Griffith, is the uptake of DNA from one organism by another organism and its ultimate expression. In this situation, *Neurospora* has been pretreated in such a way as to cause the uptake of the bacterial plasmid. This is a well-used technique in molecular genetics as a way of introducing genes from virtually any source into the organisms under study.

f. Plant and fungal cells are generally prepared for transformation by removal of their cell walls. The cell membranes are then exposed to a high salt concentration and the exogenous DNA. Studies indicate that the DNA enters the cell in two ways: (1) phagocytosis and (2) localized temporary dissolving of the membrane by the high salt concentration.

g. With successful transformation, the exogenous DNA passes through the cytoplasm and enters the nucleus, where it becomes integrated into a host chromosome.

h.

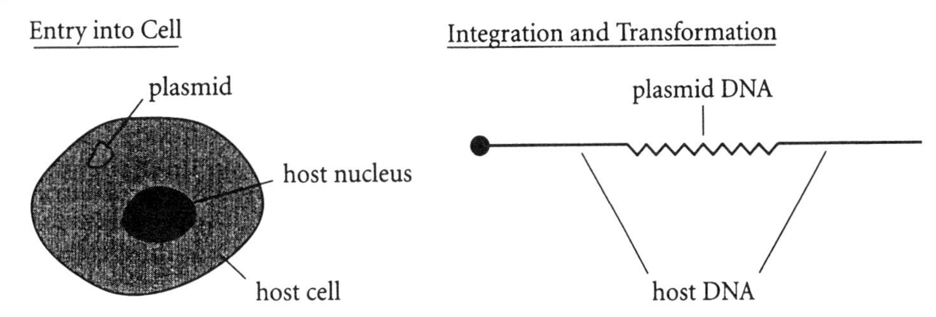

i. It is completely unnecessary to know what benomyl is. Its use simply allows for the selection of cells that received and integrated some exogenous DNA. Virtually any resistance marker could have been used. The choice of a resistance marker usually depends on what is easily available to the researcher, although questions of toxicity to humans may play a role in the choice.

j. Because hyphal extension occurs in colonies, not at the one-cell stage, the researcher must look for mutants that are expressed by a clone or colony. Therefore, he is looking for mutants that are "colonial." Mutations that produce an aberrant colony in size or shape are, by definition, involved with the extension of hyphae.

k. The "previous mutational analysis" could have been any random study. For example, in screens for specific auxotrophic mutants experimenters would have noticed this abnormal phenotype also appearing.

l.

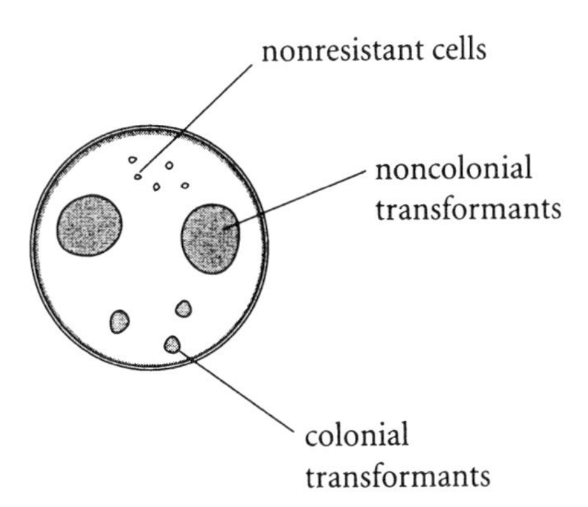

m. Tagging is a process to mutate and "mark" genes of interest by insertion of known DNA. In this case, the gene for benomyl resistance is being inserted randomly within the genome. Occasionally, insertion will occur within a gene involved with hyphal extension. These will cause the aberrant phenotype because the gene has been disrupted by the insertion of the selectable DNA. The disruption causes a knockout mutation within the gene of interest and also supplies a "molecular handle" to later clone the DNA.

n. The orange-colored bread mold *Neurospora* is a multicellular haploid in which the cells are joined end to end to form hyphae. The hyphae grow through the substrate and also send up aerial branches that bud off haploid cells known as conidia (asexual spores). Conidia can detach and disperse to form new colonies, or alternatively they can act as paternal gametes and fuse with a maternal structure of a different individual. However, the different individual must be of the opposite mating type. In *Neurospora* there are two mating types, determined by the alleles *A* and *a*. A cross will succeed only if it is $A \times a$. An asexual spore from the opposite mating type fuses with a receptive hair, and a nucleus travels down the hair to pair with a maternal gamete that waits inside a specialized knot of hyphae. The *A* and *a* pair then undergo synchronous mitoses, finally fusing to form diploid meiocytes. Meiosis occurs, and in each meiocyte four haploid nuclei are produced, which represent the four products of meiosis. For an unknown reason, these four nuclei divide mitotically, resulting in eight nuclei, which develop into eight football-shaped sexual spores called ascospores. The ascospores are shot out of a flask-shaped fruiting body that has developed from the knot of hyphae that originally contained the maternal gametic cell. The ascospores can be isolated, each into a culture tube, where each ascospore will grow into a new culture by mitosis.

o. In order for the benomyl-resistance gene to be integrated within the host chromosome, it recombines with it. However, it is recombination between the *col* and ben^R genes that is interesting. It is either 0% (type 1), indicating that the insertion caused the small-colony phenotype, or 50% (type 2), showing the two events are unlinked.

p. Only two types are possible: integration into a "hyphal" gene (so the resistance and small-colony phenotype are linked) and ectopic integration and concurrent mutation of a gene causing the small-colony phenotype where the two are unrelated and also unlinked. Which type is more likely depends on mutation rates and the number of genes that can mutate to the small-colony phenotype (i.e., the ease of generating spontaneous *col* mutants) and on the rate of transformation and integration.

q. A probe in experiments such as this one is usually a sequence of DNA that can be used to identify a specific DNA sequence within a genome or colony. The probe is labeled in some way to indicate its presence. In this experiment, the probe was probably the bacterial plasmid (although it might have been the benomyl-resistance gene), most likely radioactively labeled. A genomic library from each *col* mutant would be screened with the probe to identify those clones that contain complementary sequences (and, with luck, some sequences of the gene of interest).

r. A probe specific to the bacterial plasmid could be made by growing bacteria with the plasmid. The plasmids could be isolated through cesium chloride centrifugation and then labeled.

Solution to the Problem

a. Type 1 isolates behave genetically as if the benomyl-resistance and small-colony phenotype are completely linked. The progeny are all parental in phenotype (either *col ben-R* or + *ben-S*). This would be the expected result if the insertion of the plasmid (and *ben-R* gene) caused the mutation that led to the *col* phenotype, which of course was the point of the experiment.

Type 2 isolates behave genetically as if the benomyl-resistance and small-colony phenotype are unlinked (i.e., the two markers are segregating randomly in the progeny). This is the expected result if the insertion of the *ben-R* gene was unrelated to the mutation that led to the *col* phenotype. In other words, the *col* mutation occurred randomly and separately from the insertion of the *ben-R* gene.

b. The type 1 isolates are the *col* genes that are mutated by the insertion of the plasmid, and therefore these are the genes that are tagged.

c. The type 1 isolates should be used to create genomic libraries. The libraries should be screened for clones that contain DNA adjacent to the insert by probing with known sequences from the plasmid. The identified clones will represent parts of the disrupted gene of interest. To recover the intact wild-type gene, a subclone of the disrupted gene sequence can be used to probe a wild-type genomic library.

d. All progeny that are benomyl-resistant will also contain DNA from the bacterial plasmid integrated into their chromosome or chromosomes. Thus, all *ben-R* strains from this experiment will have DNA that hybridizes to a probe specific for the plasmid.

12 Genomics

1. ***Unpacking the Problem***

   a. Two types of hybridizations that have already been discussed are hybridizations between strains of a species and hybridizations between species. A third type of hybridization is referred to in this problem: molecular hybridization. Molecular hybridization can involve either DNA-DNA hybridization or DNA-RNA hybridization. In both instances, it relies on the specificity of complementary pairing and can take place in solution, on a gel, on a filter, or on a slide. For example:

   5′- U A C G G G A U - 3′ RNA
   3′- A T G C C C T A - 5′ DNA

   b. In situ hybridization usually is conducted on a slide so that the stained chromosomes can be observed and the specific portion of a chromosome to which the probe hybridizes can be identified.

   c. A YAC is a yeast artificial chromosome. It contains a yeast centromere, autonomous replication sequences (origins of replication), telomeres, and DNA that has been attached between them.

   d. Chromosome bands are dark regions along the length of a chromosome that occur in a characteristic pattern for each chromosome within an organism. They can occur naturally, as with *Drosophila* polytene chromosomes, or they can be induced by a number of chemical and physical agents, combined with staining to accentuate the bands and interbands.

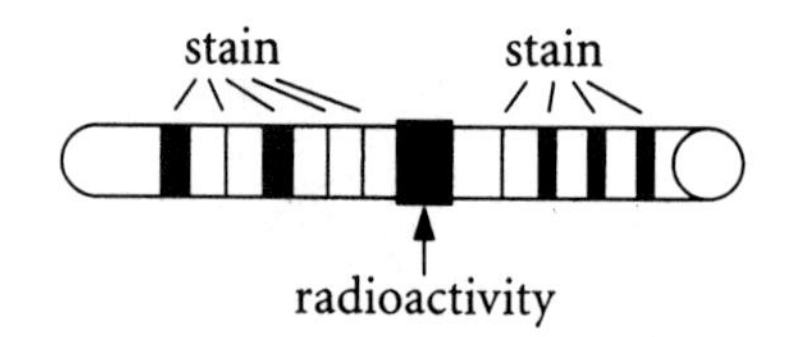

e. The five YACs could have been hybridized sequentially to the same chromosome preparation, which is, however, unlikely. Alternatively, the information could have been determined in five separate experiments. In either case, a YAC labeled with either radioactivity or fluorescence, and including the DNA of interest, was hybridized to a chromosome preparation. The chromosomes were properly treated to reveal the banding pattern, and the YACs were determined to hybridize to the same band.

f. A genomic fragment, by definition, contains a subportion of the genome being studied. In most instances, it actually contains a subportion of one chromosome. Five randomly chosen YACs would not be expected to contain the same genomic fragment or even fragments from the same chromosome. The fragments could have been produced by either physical (X-irradiation, shearing) or chemical (digestion, restriction) means, but it does not matter how they were produced.

g. A restriction enzyme is a naturally occurring bacterial enzyme that is capable of causing either single- or double-stranded breaks in DNA at specific DNA sequences.

h. A long cutter is a restriction enzyme that produces very long fragments of DNA because the sequence it recognizes occurs infrequently within the genome.

i. The YACs were radioactively labeled so that their location after hybridization could be detected through autoradiography. To radioactively label is to attach an isotope that emits energy through decay. Commonly used radioactive labels are tritium (^3H) in place of hydrogen and ^{32}P in place of phosphorus.

j. An autoradiogram is a "self-picture" taken through radioactive decay from a labeled probe. When a gel or blot is used, the radioactive decay is captured by a piece of X-ray film. When in situ hybridization is performed on slides, the photographic emulsion coats the slide directly.

k. Free choice. Be sure you truly know the meaning of each term.

l. We are given a diagram of the composite autoradiographic results. The DNA from humans was isolated and subjected to digestion by a restriction enzyme that cuts very infrequently. Once the DNA was electrophoresed, it was Southern-blotted and then probed sequentially with radioactively labeled YACs, followed by sequential exposure to X-ray film. Between probings, the previous YAC hybrid was removed through denaturing of the DNA-DNA hybrid. Alternatively, five separate Southern blottings were done.

m. The haploid human genome is thought to contain approximately 3.3×10^6 kilobases of DNA.

n. Restriction digestion of human genomic DNA would be expected to produce hundreds of thousands of fragments.

o. The fragments produced by restriction of human genomic DNA would be expected to be mostly different.

p. When subjected to electrophoresis and then stained with a DNA stain, the digested human genome would produce a continuous "smear" of DNA, from very large fragments (in excess of tens of thousands of base pairs in length) to fragments that are very small (under a hundred bases in length).

q. In this question, only two distinct bands are produced, at most, in any one probing. The difference between what is seen with a DNA stain and what is seen with probing lies in the specificity of the agent being used. DNA stain will detect any DNA, while a DNA probe will detect only DNA that is complementary to the probe.

r. Number them from top to bottom, 1–3, across the gel. Thus, YACs A–C contain band 1, YACs C–D contain band 2, and YACs A and E contain band 3.

s. There are no restriction fragments on the autoradiogram. The fragments are on the filter (nitrocellulose, nylon) used to blot the gel. The radioactivity of the probes is captured by the X-ray film as it decays, producing an exposed region of film.

t. YACs B, D, and E hybridize to one fragment, and YACs A and C hybridize to two fragments.

u. A YAC can hybridize to two fragments if the YAC contains continuous DNA and there is a restriction site within that region. A YAC can also hybridize to two fragments if it contains discontinuous DNA from two locations in the genome that either are on different chromosomes (this is analogous to a translocation) or are separated by at least two restriction sites if they are on the same chromosome (this is analogous to a deletion). In this case, the former makes more sense. Because the YACs were selected for their binding to one specific chromosome band, it is unlikely that the YACs are composed of discontinuous DNA sequences. A YAC could hybridize to more than two fragments because the continuous DNA could contain many restriction sites or the discontinuous DNA could be composed of DNA from a number of regions in the genome.

v. Cytogeneticists use the term *band* to designate a region of a chromosome that is dark-staining. Molecular biologists use *band* to designate a region of dark appearing on an autoradiogram, which is produced by radioactive decay from a specific probe that reacted with a population of molecules localized by gel electrophoresis. In both cases, *band* refers to a localization.

Solution to the Problem

a. Note that fragments 1 and 3 occur together and fragments 1 and 2 occur together, but that fragments 2 and 3 do not occur together. This suggests that the sequence is 2 1 3 (or 3 1 2).

b. If the sequence of the fragments is 2 1 3, then the YACs can be shown in relation to these fragments. YAC A spans at least a portion of both 1 and 3. YAC B is within region 1. YAC C spans at least a portion of regions 1 and 2. YAC D is contained within region 2. YAC E is contained within region 3. A diagram of these results is shown below. In the diagram, there is no way to know the exact location of the ends of each YAC.

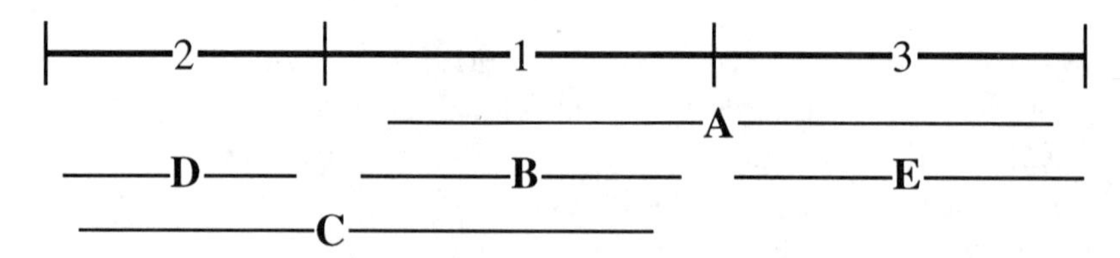

2. PFGE results in separation of the chromosomes, and the gel indicates the discreteness obtained by the separation. Because the *mata* probe does not hybridize to a band, there must not be a gene of similar sequence in *Neurospora.* On the other hand, the *leu2* probe hybridizes to a sequence contained within chromosome 4, and *ade3* to a sequence contained within chromosome 7.

3. The cross is

$$cys\text{-}1\ RFLP\text{-}1^{O}\ RFLP\text{-}2^{O} \times cys\text{-}1^{+}\ RFLP\text{-}1^{M}\ RFLP\text{-}2^{M}$$

Scoring the progeny, a parental type will have the genotype of either strain and, if the markers are all linked, be the most common. A recombinant type will have a mixed genotype and be less common. Clearly, the first two ascospore types are parental, with the remaining being recombinant.

a. The *cys-1* locus is in this region of chromosome 5. If it were not in this region, linkage to either of the RFLP loci would not be observed.

b. To calculate specific distances, you may need to review previous chapters. Here, it is assumed that you recall basic mapping strategies.

$$cys\text{-}1 \text{ to } RFLP\text{-}1 = {}^{(2\,+\,3)}/_{100} \times 100\% = 5 \text{ map units}$$
$$cys\text{-}1 \text{ to } RFLP\text{-}2 = {}^{(7\,+\,5)}/_{100} \times 100\% = 12 \text{ map units}$$
$$RFLP\text{-}1 \text{ to } RFLP\text{-}2 = {}^{(2\,+\,3\,+\,7\,+\,5)}/_{100} \times 100\% = 17 \text{ map units}$$

|—5 m.u.—|—————12 m.u.—————|

RFLP-1   *cys-1*   RFLP-2

c. A number of strategies could be tried. Since this is an auxotrophic mutant, functional complementation can be attempted. Positional cloning or chromosome walking from the RFLPs is also a very common strategy.

4. **a.**, **b.** Compare each translocation gel to the wild-type gel and note the difference in bands that were obtained. Together, the translocations involve chromosomes 1, 2, and 4. The top band in all three gels is constant and must reflect chromosome 3, which is not involved in either translocation. Focusing on the three remaining bands in the wild-type gel, the last band is not altered in the 1;4 translocation but is altered in the 2;4 translocation. This means that it must reflect chromosome 2. The band second from the bottom in the wild-type gel is altered in the 1;4 translocation but is not altered in the 2;4 translocation and must reflect chromosome 1. The remaining band in the wild-type gel is altered in both the 1;4 and the 2;4 translocations and it must reflect chromosome 4.

In the 1;4 lane, the two new bands in comparison with the wild-type lane are due to the 1;4 translocation, while in the 2;4 lane, they are due to the 2;4 translocation. In both translocation lanes, the P probe is associated with new-appearing bands, indicating that P is located on chromosome 4. This is confirmed in the wild-type lane.

5. Remember that a gene is one small region of a long strand of DNA and that a cloned gene will contain the entire sequence of the gene under normal circumstances. If there are two cuts within a gene, three fragments will be produced, all of which will interact with the probe, as was seen with enzyme 1. Cuts external to a gene will produce one fragment that will interact with the probe, which was seen with enzyme 2. One cut within a gene will produce two fragments that will interact with the probe, as was seen with enzyme 3.

6. **a.** To determine the physical map showing the STS order, simply list the STSs that are positive, using parentheses if the order is unknown, and align them with one another to form a consistent order.

YAC A:			1	4	3	
YAC B:		5	1			
YAC C:				4	3	7
YAC D:	(6 2)	5				
YAC E:					3	7

b. Once the sequence of STSs is known, the YACs can be aligned as follows, although precise details of overlapping and the locations of ends are unknown:

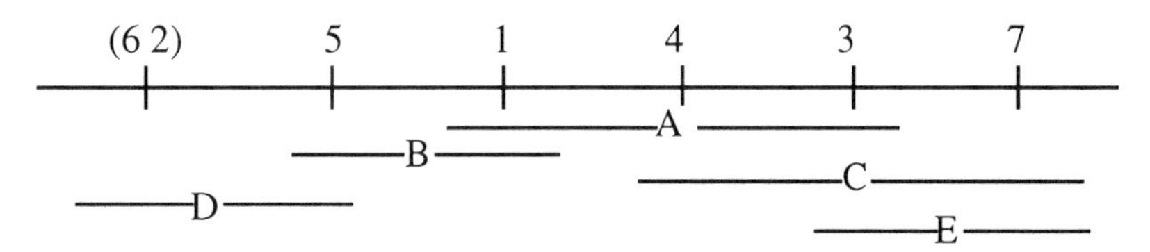

7. **a., b.** There are four patterns that can be observed in the comparisons that can be made between these six markers: + +, – –, + –, and – +. The first two indicate concordance and the second two indicate a lack of concordance. Ideally, data would show either 100% concordance for the seven hybrids, indicating linkage, or 100% discordance for the seven hybrids, indicating a lack of linkage.

Because radiation hybrids involve chromosome breakage, two genes that are located very close together on the same chromosome may show some discordance despite the close linkage. Two genes that are located on different chromosomes may also show some concordance due to the chance that two separate fragments may become established within a single hybrid line. Therefore, the problem is how to distinguish between reduced concordance due to chromosome fragmentation and chance concordance due to two fragments from different chromosomes being in the same hybrid. Obviously, a statistical solution is needed, but there are not enough data in this problem for a statistical analysis.

Sort the data into three groups: 100% concordance, 100% discordance, and mixed (concordance/discordance). This follows below:

100% Concordance	100% Discordance	Mixed
E-F	None	A-B 2/5
		A-C 2/5
		A-D 6/1
		A-E 2/5
		A-F 2/5
		B-C 5/2
		B-D 1/6
		B-E 3/4
		B-F 3/4
		C-D 3/4
		C-E 3/4
		C-F 3/4
		D-E 3/4
		D-F 3/4

Markers E and F are most likely located on the same chromosome. Markers B and D may be located on different chromosomes.

In the absence of statistical analysis, with so few total hybrids, it is important to pay more attention to the + + patterns than the – – patterns simply because – – can arise either from linkage, with the specific chromosome missing in the hybrid, or from lack of linkage, with the two chromosomes (or fragments) lacking in the hybrid. Therefore, going back to the mixed category and focusing on those marker pairs that had a high degree, but not 100%, of concordance, one sees that the 6/1 pattern of A-D and the 5/2 pattern of B-C stand out. For the A-D pair, 3 of the 6 concordances are + +, while only 2 of the 5 concordances for B-C are + +. It is unclear from the data whether this is a significant difference, and significance cannot be determined in any fashion. Therefore, it would be important to collect more data before drawing further conclusions.

8. a. RAPDs are formed when regions of DNA are bracketed by two inverted copies of a "random" PCR primer sequence. Below, the primer is indicated by X's, and the amplified region appears in brackets. For convenience, the two amplified regions are shown on the same lengthy piece of DNA for strain 1.

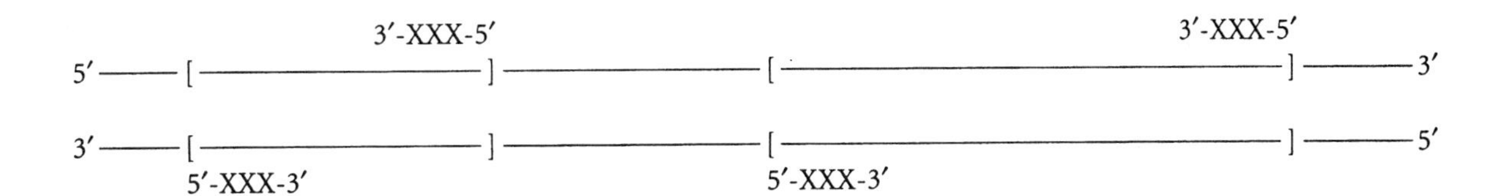

Strain 2 lacks one or two regions complementary to the primer.

b. Progeny 1 and 6 are identical with the strain 1 parent. Progeny 4 and 7 are identical with the strain 2 parent. Progeny 2 and 5 received the chromosome holding the upper band from the strain 1 parent and the chromosome holding the lower band from the strain 2 parent (resulting in no second band). Progeny 3 received the opposite: the chromosome holding the lower band from the strain 1 parent and the chromosome holding the upper band from the strain 2 parent (resulting in no second band). Therefore, bands 1 and 2 appear to be unlinked.

c. Recall that a nonparental ditype has two types only, both of which are recombinant. Therefore, the tetrad would be composed of two progeny like progeny 2 and two progeny like progeny 3.

9. **a.** Of the regions of overlap for cosmids C, D, and E, region 5 is the only region in common. Thus, gene x is localized to region 5.

b. The common region of cosmids E and F, or the location of gene y, is region 8.

c. Both probes are able to hybridize with cosmid E because the cosmid is long enough to contain part of genes x and y.

10. α: The only chromosome missing in A and B and present in C is 7.
β: The only chromosome present in all colonies is 1.
γ: The only chromosome missing in A and present in B and C is 5.
δ: The only chromosome present only in B is 6.
ε: Not on chromosomes 1 through 8.

11. For each enzyme, the goal is to determine which chromosome or chromosome arm is found in all positive cell lines and is also absent in all negative cell lines. The data indicate the following gene locations:

Steroid sulfatase: Xp
Phosphoglucomutase-3: 6q
Esterase D: 13q
Phosphofructokinase: 21
Amylase: 1p
Galactokinase: 17q.

12. **a.** The following stylized schematic of a reciprocal translocation between chromosome 3 and 21 is arbitrarily chosen to show the salient details. Band 3.1 of the q arm of chromosome 3 is split by the translocations that are correlated to the *N* disease allele. Probe c hybridizes to the region of 3q3.1 that remains with chromosome 3 and probes a, b, and d hybridize to the region of 3q3.1 that is translocated in this case to chromosome 21.

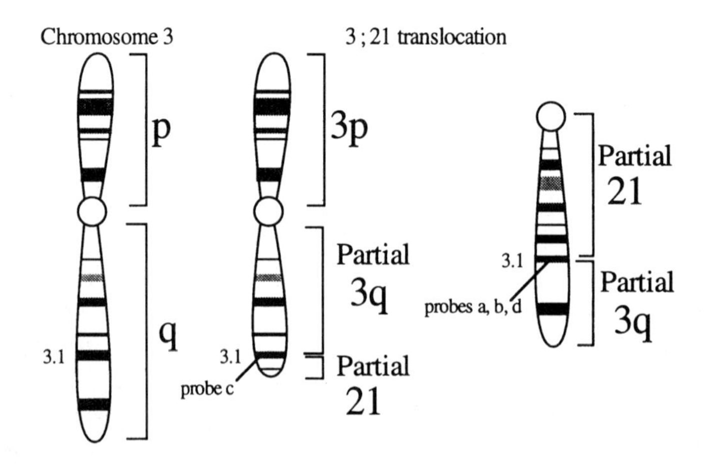

b. Since translocations of chromosome 3 that break band 3q3.1 are correlated to the disease, it is reasonable to assume that these rearrangements split the normal gene (*n*) in two, separating vital coding or regulatory regions. Therefore analysis and cloning of this specific region should be attempted.

In order to isolate and characterize the normal allele, chromosome walking from the known clones should be attempted in genomic libraries from individuals with the translocation and affected with the disease. Probe c is to one side of the breakpoint, while a, b, and d are on the other side. Also, translocation breakpoints serve as useful molecular landmarks, since they are easily identified on Southerns as "split bands" when probed with cloned DNA spanning the breakpoint. Once the breakpoint has been identified and cloned, the appropriate subclones would be used to clone the normal allele from a "normal" genomic library. This would be in conjunction with the usual techniques to identify a gene: sequencing, open reading frame analysis, Northern blots, etc.

c. Once *n* is cloned, it can be used to clone the various alleles from individuals who have the disease but not a translocation. The various alleles could then be compared with *n* by sequence, regulation, etc.

13. a. DNA from each individual was obtained. It was restricted, electrophoresed, blotted, and then probed with the five probes. After each probing, an autoradiograph was produced.

b. First identify which chromosome came from the affected parent. This is easily determined by identifying which chromosome could not have come from the mother. For the first daughter, the chromosome with 2′ was inherited from the father. Likewise, 2″, 3″, and 2″ identify the paternal chromosome in the other children. In all cases, the chromosome drawn to the left is the one inherited from the mother.

Next compare the maternal chromosomes of affected offspring with unaffected offspring to determine which RFLP is most closely correlated to the disease. This analysis is based on the co-segregation of one of the RFLPs and the disease-causing gene. Notice that all of these chromosomes show evidence of recombination. For example, when compared with the mother's chromosomes, it can be deduced that the maternally inherited chromosome of the unaffected daughter is the result of a double crossover event.

Affected:	$1^o\ 2^o\ 3'\ 4^o\ 5^o$
Unaffected:	$1^o\ 2^o\ 3'\ 4'\ 5^o$
Unaffected:	$1'\ 2^o\ 3'\ 4'\ 5^o$
Affected:	$1'\ 2^o\ 3'\ 4^o\ 5'$

The only RFLP that correlates to the disease and therefore is likely closest to the disease allele is 4°. It is present in both affected children and absent in both unaffected children.

c. It appears that RFLP 4 is the closest marker to the gene and could be used for positional cloning by chromosome walking. However, with only four offspring, the genetic distance between the gene and this marker could be quite large. The number of markers for each human chromosome is already large and increasing almost daily. If possible, it makes sense to further analyze this family (and as many other families with the same trait that can be found) to see if the gene can be further localized before the arduous task of "walking" is attempted.

14. a. Breaks in different regions of 17R result in deletion of all genes from the breakpoint.

b. Because the only human DNA is from 17R, all the human genes expressed must be on 17R. Notice that only gene *c* is expressed by itself. This means that gene *c* is closest to the mouse DNA. Next notice that if *c* and one other gene are expressed, that other gene is always *b*. This means *b* is closer to the mouse DNA than *a* is. The gene order is mouse—*c*—*b*—*a*.

The probability of a break between two genes is a function of the distance between them. Of the 200 lines tested, 48 expressed no human activity. Thus, the breakpoint was between the *c* gene and the mouse centromere. Similarly, a break between *c* and *b* (cells express c only) occurred in 12 lines, and a break between *b* and *a* (cells express c and b) occurred in 80 lines. Finally, 60 lines expressed all three genes, placing the breakpoint between *a* and the end of the chromosome.

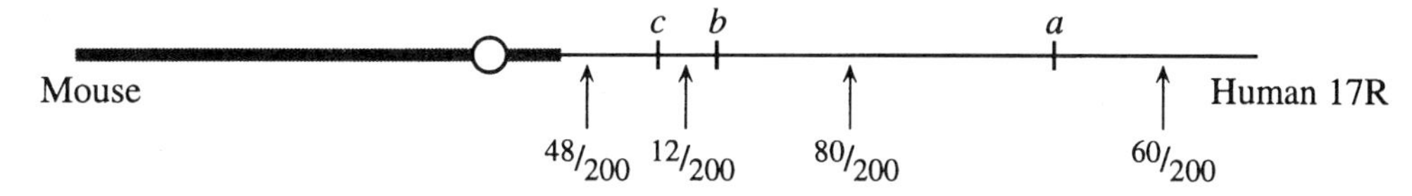

c. The banding patterns of 17R could be used to verify the suggested map as well as provide a more accurate position for the three genes. For example, the 60 lines that express all three genes could be analyzed to see how much of 17R can be removed without deleting gene *a*. The line with the most removed will have the closest breakpoint to the gene from the end. By analyzing the 80 lines that express *c* and *b* but not *a*, the line with the least deleted will define the breakpoint that is closest to *a* from the centromere. By comparing both, the region in between defines the location of gene *a*.

15. a., b., c. Cystic fibrosis (CF) is a recessive, autosomally inherited disease. Both parents in this pedigree must be carriers since some of their children are affected. Since the problem states that the three probes used are very closely linked to the *CF* gene, recombination will be ignored.

The data from three probes are presented, but only probes 1 and 3 detect RFLPs in this pedigree and are therefore informative. Both probes detect either one or two bands depending on the allele present. Calling the one-band pattern allele *A* and the two-band pattern allele *B*, the individuals of the pedigree are

Father	$RFLP\text{-}1^B$ $RFLP\text{-}3^A$
Mother	$RFLP\text{-}1^A$ $RFLP\text{-}3^B$
Child 1 (II-1)	$RFLP\text{-}1^B$ $RFLP\text{-}3^A$ (does not have CF)
Child 2 (II-2)	$RFLP\text{-}1^B$ $RFLP\text{-}3^B$ (does have CF)
Child 3 (II-3)	$RFLP\text{-}1^B$ $RFLP\text{-}3^B$ (does have CF)
Child 4 (II-4)	$RFLP\text{-}1^A$ $RFLP\text{-}3^B$ (does not have CF)
Child 5 (II-5)	$RFLP\text{-}1^B$ $RFLP\text{-}3^A$ (does not have CF)
Child 6 (II-6)	$RFLP\text{-}1^B$ $RFLP\text{-}3^B$ (does have CF)
Child 7 (II-7)	$RFLP\text{-}1^A$ $RFLP\text{-}3^A$ (does not have CF)

The first step is to determine which RFLP alleles are linked to the disease-causing *CF* alleles. The pattern of inheritance suggests that $RFLP\text{-}1^B$ from the father and $RFLP\text{-}3^B$ from the mother are both linked to *CF* alleles since all children that are $RFLP\text{-}1^B$ $RFLP\text{-}3^B$ also have CF.

The oldest son (II-1) is a carrier since he has inherited a *CF* allele (linked to $RFLP\text{-}1^B$) from his father. Similarly, II-4 has inherited a *CF* allele from his mother, II-5 has inherited a *CF* allele from his father, and II-7 is homozygous normal.

13 Transposable Genetic Elements

1. Mutations in *gal* can be generated and from these strains λ*dgal* phage isolated. Through hybridization of denatured λ*dgal* DNA containing the mutation with wild-type λ*dgal* DNA, some of the molecules will be heteroduplexes between one mutant and one wild-type strand. If the mutation was caused by an insertion, the heteroduplexes will show a "looped out" section of single-stranded DNA, confirming that one DNA strand contains a sequence of DNA not present in the other (see Figure 13-2 in the companion text).

   Text Figure 13-1 illustrates a method to compare the densities of gal^{+}-carrying λ phage with gal^{-}-carrying phage. In this experiment, the gal^{-}-phage are denser, indicating that they contain a larger DNA molecule.

   If the *gal* genes are cloned, direct comparison of the restriction maps or even the DNA sequence of mutants compared with wild type will give specific information about whether any are the results of insertions.

2. In replicative transposition, transposable elements move to a new location by replicating into the target DNA, leaving behind a copy of the transposable element at the original site. If, on the other hand, the transposable element excises from its original position and inserts into a new position, this is called **conservative transposition.**

   To test either mechanism, experiments must be designed so that both the "old" and "new" positions of the transposon can be assayed. If the transposon remains in the old site at the same time that a new copy is detected elsewhere, a replicative mechanism must be in use. If the transposon no longer exists in the old site when a copy is detected elsewhere, a conservative mechanism must be in use.

Figure 13-12 in the companion text describes how replicative transposition can be observed between two plasmids. The same general protocol could be used to detect conservative transposition, but of course the results would be different.

Kleckner and co-workers actually demonstrated conservative transposition by following the movement of a transposon that contained a small heteroduplex within the *lacZ* gene. The DNA of two derivatives of Tn10 carrying different *lacZ* alleles (one being wild-type and the other being mutant) were denatured and allowed to reanneal. In some cases, the DNA molecules that reformed were actually heteroduplexes; one strand contained the *lacZ*$^+$ allele, and the other strand contained the *lacZ*$^-$ allele. Transpositions of theses heteroduplexes were then followed. Based on the mechanism of movement, two outcomes are possible. If replicative transposition occurred, the semiconservative nature of DNA replication would generate two genetically different transposons: instead of the heteroduplex *lacZ*$^+$/*lacZ*$^-$ DNA, one would now be all *lacZ*$^+$ and the other all *lacZ*$^-$. If transposition was conservative, the *lacZ* gene would still be heteroduplex *lacZ*$^+$/*lacZ*$^-$ after transposition and the first cell division would resolve the heteroduplex. This is what was observed. (The "sectored" colonies are the result of the original cell still having the *lacZ*$^+$/*lacZ*$^-$ heteroduplex after transposition. After the first division one cell was now *lacZ*$^+$ and the other was *lacZ*$^-$.) The experiment is outlined below:

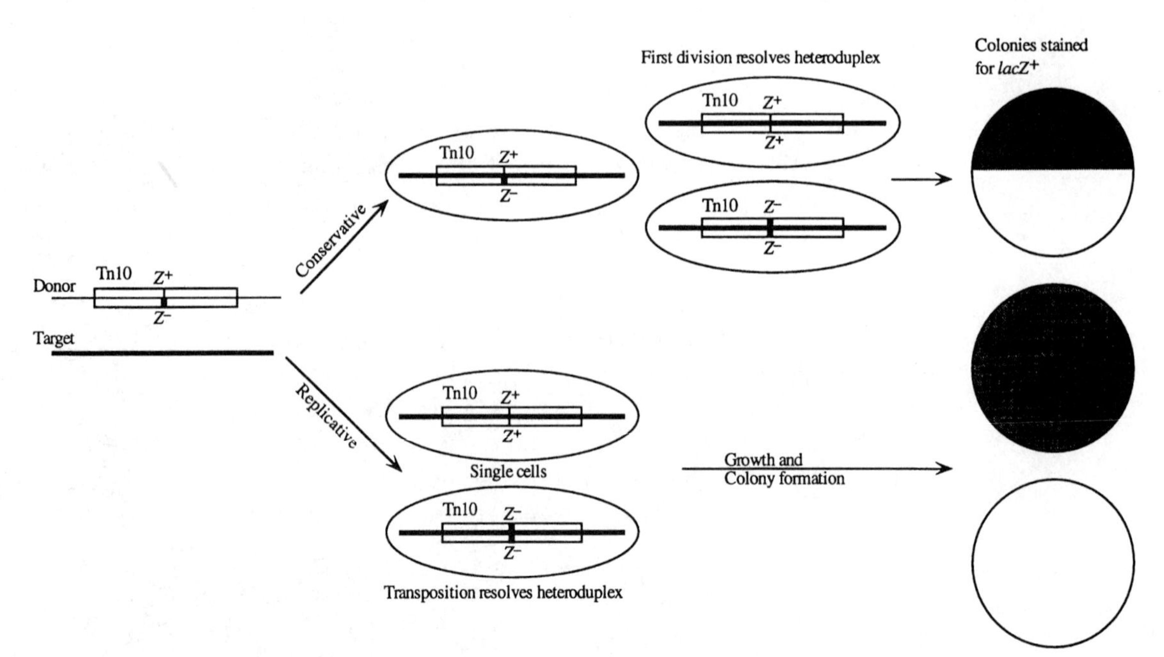

3. R plasmids are the main carriers of drug resistance. These plasmids are self-replicating and contain any number of genes for drug resistance as well as the genes necessary for transfer by conjugation (called the **RTF region**). It is the R plasmid's ability to transfer rapidly to other cells, even those of related species, that allows drug resistance to spread so rapidly. R plasmids acquire drug-resistance genes through transposition. Drug-resistance genes are found flanked by IR (inverted repeat) sequences and as a unit are known as **transposons.** Many transposons have been identified, and as a set they encode a wide range of drug resistances (see Table 13-3 in the text). Since transposons can "jump" between DNA molecules (e.g., from one plasmid to another or from a plasmid to the bacterial chromosome and vice versa), plasmids can continue to gain new drug-resistance genes as they mix and spread through different strains of cells. It is a classic example of evolution through natural selection. Those cells harboring R plasmids with multiple drug resistances survive to reproduce in the new environment of antibiotic use.

4. Boeke, Fink, and their co-workers demonstrated that transposition of the Ty element in yeast involved an RNA intermediate. They constructed a plasmid using a Ty element that had a promoter that could be activated by galactose, and an intron inserted into its coding region. First, the frequency of transposition was greatly increased by the addition of galactose, indicating that an increase in transcription (and production of RNA) was correlated to rates of transposition. More importantly, after transposition they found that the newly transposed Ty DNA lacked the intron sequence. Because intron splicing occurs only during RNA processing, there must have been an RNA intermediate in the transposition event.

5. P elements are transposable elements found in *Drosophila.* Under certain conditions they are highly mobile and can be used to generate new mutations by random insertion and gene knockout. As such, they are a valuable tool to tag and then clone any number of genes. (See Problem 15 from Chapter 11 for a discussion on cloning by tagging.) P elements can also be manipulated and used to insert almost any DNA (or gene) into the *Drosophila* genome. P element–mediated gene transfer works by inserting the DNA of interest between the inverted repeats necessary for P element transposition and injecting this recombinant DNA along with helper intact P element DNA (to supply the transposase) into very early embryos and screening for (random) insertion among the injected fly's offspring.

6. The a_1 allele is an unstable mutant allele. It contains a "defective" transposable element that requires a trans-factor produced by the unlinked *Dt* locus to move (and revert a_1 to A_1, restoring its function). In other words, the a_1 allele is nonautonomous and requires *Dt* to revert. Since Rhoades chose pollen from fully pigmented anthers, the tissue had already undergone a reversion event and was likely A_1/a_1 ; *Dt/Dt.* The pollen from these anthers would be A_1 ; *Dt* and a_1 ; *Dt.* Since each kernel is the result of a separate fertilization, some will be A_1/a_1 ; *Dt/dt* (and fully pigmented) and others will be a_1/a_1 ; *Dt/dt.* The latter can still undergo reversion events during somatic growth, resulting in "dotted" kernels. Each dot would represent the descendants of a cell that had undergone a similar excision–reversion event.

7. The sn^+ patches in an *sn* background and the occurrence of sn^+ progeny from an $sn \times sn$ mating indicate that sn^+ alleles are appearing at relatively high frequencies and that the *sn* alleles are unstable. High reversion rates suggest that the *sn* allele is the result of an insertion of a transposable element that is capable of (frequent) excision.

8. **a.**, **b.** The soil bacterium *Agrobacterium tumefaciens* contains a large plasmid called the **Ti (tumor-inducing) plasmid.** When this bacterium

infects a plant, a region of the Ti plasmid called the **T-DNA** is transferred and inserted randomly into the plant's genome. The T-DNA directs the synthesis of plant hormones that cause uncontrolled growth (a tumor) and also directs the plant's synthesis of compounds called **opines.** (These compounds cannot be metabolized by the plant but are used by the bacterium.)

When a piece of "normal" plant tissue is cultured with appropriate nutrients and growth hormones, cells are stimulated to divide in a disorganized manner, forming a mass of undifferentiated cells called a **callus.** These cells will differentiate only into shoots (or roots) if the levels of growth hormones are carefully adjusted. The T-DNA causes undifferentiated growth because it directs the unbalanced synthesis of these same hormones. The fact that some of the infected cultures produced shoots suggests that these cells "lost" the ability to overproduce these hormones. This would be consistent with the loss of the T-DNA (similar to the loss of other transposable elements that is observed in many species). Thus, the A graft would grow normally, and seeds produced by the graft would have no trace of the T-DNA. The fact that cells from the A graft grow like tumor cells when placed on synthetic medium suggests that the medium supplies the high levels of hormones necessary for undifferentiated growth even in the absence of T-DNA.

9. In the *Ac-Ds* system, *Ac* can produce an unstable allele that is autonomous. *Ds* can revert only in the presence of *Ac* and is nonautonomous. In the following, Ac^+ indicates the absence of the *Ac* regulator gene.

Cross 1:

P $C/c^{Ds}; Ac/Ac^+ \times c/c; Ac^+/Ac^+$

F_1
- $^1/_4$ $C/c; Ac/Ac^+$ (solid pigment)
- $^1/_4$ $C/c; Ac^+/Ac^+$ (solid pigment)
- $^1/_4$ $c^{Ds}/c; Ac/Ac^+$ (unstable colorless or spotted)
- $^1/_4$ $c^{Ds}/c; Ac^+/Ac^+$ (colorless)

Overall: 2 solid:1 spotted:1 colorless

Cross 2:

P $C/c^{Ac} \times c/c$

F_1
- $^1/_2$ C/c (solid pigment)
- $^1/_2$ c/c^{Ac} (unstable colorless or spotted)

Overall: 1 solid:1 spotted

Cross 3:

P $C/c^{Ds}; Ac/Ac^+ \times C/c^{Ac}; Ac^+/Ac^+$

F_1
- $^1/_8$ $C/C; Ac/Ac^+$ (solid pigment)
- $^1/_8$ $C/c^{Ac}; Ac/Ac^+$ (solid pigment)
- $^1/_8$ $C/C; Ac^+/Ac^+$ (solid pigment)
- $^1/_8$ $C/c^{Ac}; Ac^+/Ac^+$ (solid pigment)
- $^1/_8$ $C/c^{Ds}; Ac^+/Ac^+$ (solid pigment)
- $^1/_8$ $C/c^{Ds}; Ac^+/Ac$ (solid pigment)
- $^1/_8$ $c^{Ds}/c^{Ac}; Ac^+/Ac^+$ (unstable colorless or spotted)
- $^1/_8$ $c^{Ds}/c^{Ac}; Ac^+/Ac$ (unstable colorless or spotted)

Overall: 3 solid:1 spotted

14 Regulation of Gene Transcription

1. The I gene determines the synthesis of a repressor molecule, which blocks expression of the *lac* operon and which is inactivated by the inducer. The presence of the repressor I^+ will be dominant to the absence of a repressor I^-. I^S mutants are unresponsive to an inducer. For this reason, the gene product cannot be stopped from interacting with the operator and blocking the *lac* operon. Therefore, I^S is dominant to I^+.

2. O^c mutants are changes in the DNA sequence of the operator that impair the binding of the *lac* repressor. Therefore, the *lac* operon associated with the O^c operator cannot be turned off. Because an operator controls only the genes on the same DNA strand, it is cis (on the same strand) and dominant (cannot be turned off).

3. **a.** You are told that a, b, and c represent *lacI*, *lacO*, and *lacZ*, but you do not know which is which. Both a^- and c^- have constitutive phenotypes (lines 1 and 2) and therefore must represent mutations in either the operator (*lacO*) or the repressor (*lac I*). b^- (line 3) shows no ß-gal activity and by elimination must represent the *lacZ* gene.

   Mutations in the operator will be cis-dominant and will cause constitutive expression of the *lacZ* gene only if it's on the same chromosome. Line 6 has c^- on the same chromosome as b^+ but the phenotype is still inducible (owing to c^+ in trans). Line 7 has a^- on the same chromosome as b^+ and is constitutive even though the other chromosome is a^+. Therefore a is *lacO*, c is *lacI*, and b is *lacZ*.

   b. Another way of labeling mutants of the operator is to denote that they lead to a constitutive phenotype; *lacO*$^-$ (or a^-) can also be written as *lacO*^{c}.

There are also mutations of the repressor that fail to bind inducer (allolactose) as opposed to fail to bind DNA. These two classes have quite different phenotypes and are distinguished by *lacI*^{S} (fails to bind allolactose and leads to a dominant uninducible phenotype in the presence of a wild-type operator) and *lacI*$^-$ (fails to bind DNA and is recessive). It is possible that line 3, line 4, and line 7 have *lacI*^{S} mutations (since dominance cannot be ascertained in a cell that is also *lacO*^{C}) but the other c^- alleles must be *lacI*$^-$.

4.

	ß-Galactosidase		**Permease**	
Part	No lactose	Lactose	No lactose	Lactose
a	+	+	–	+
b	+	+	–	–
c	–	–	–	–
d	–	–	–	–
e	+	+	+	+
f	+	+	–	–
g	–	+	–	+

5. If there is an operon governing both genes, then a frameshift mutation could cause the stop codon separating the two genes to be read as a sense codon. Therefore, the second gene product will be incorrect for almost all amino acids. However, there are no known polycistronic messages in eukaryotes. The alternative, and better, explanation is that both enzymatic functions are performed by the same gene product. Here, a frameshift mutation beyond the first function, carbamyl phosphate synthetase, will result in the second half of the protein molecule being nonfunctional.

6. Because very small amounts of the repressor are made, the system as a whole is quite responsive to changes in lactose concentration. In the heterodiploids, repressor tetramers may form by association of polypeptides encoded by I^- and I^+. The operator binding site binds two subunits at a time. Therefore, the repressors produced may reduce operator binding, which in turn would result in some expression of the *lac* genes in the absence of lactose.

7. A gene is turned off or inactivated by the "modulator" (usually called a *repressor*) in negative control, and the repressor must be removed for transcription to occur. A gene is turned on by the "modulator" (usually called an *activator*) in positive control, and the activator must be added or converted to an active form for transcription to occur.

8. The *lacY* gene produces a permease that transports lactose into the cell. A cell containing a *lacY*$^-$ mutation cannot transport lactose into the cell, so ß-galactosidase will not be induced.

9. Activation of gene expression by trans-acting factors occurs in both prokaryotes and eukaryotes. In both cases, the trans-acting factors interact with specific DNA sequences that control expression of cis genes.

In prokaryotes, proteins bind to specific DNA sequences, which in turn regulate one or more downstream genes.

In eukaryotes, highly conserved sequences such as CCAAT and various enhancers in conjunction with trans-acting binding proteins increase transcription controlled by the downstream TATA box promoter. Several proteins have been found that bind to the CCAAT sequence, upstream GC boxes, and the TATA sequence in *Drosophila*, yeast, and other organisms. Specifically, the Sp1 protein recognizes the upstream GC boxes of the SV40 promoter and many other genes; GCN4 and GAL4 proteins recognize upstream sequences in yeast; and many hormone receptors bind to specific sites on the DNA (e.g., estrogen complexed to its receptor binding to a sequence upstream of the ovalbumin gene in chicken oviduct cells). Additionally, the structure of some of these trans-acting DNA-binding proteins is quite similar to the structure of binding proteins seen in prokaryotes. Further, protein-protein interactions are important in both prokaryotes and eukaryotes. For the above reasons, eukaryotic regulation is now thought to be very close to the model for regulation of the bacterial *ara* operon.

10. Bacterial operons contain a promoter region that extends approximately 35 bases upstream of the site where transcription is initiated. Within this region is the promoter. Activators and repressors, both of which are trans-acting proteins that bind to the promoter region, regulate transcription of associated genes in cis only.

The eukaryotic gene has the same basic organization. However, the promoter region is somewhat larger. Also, enhancers up to several thousand nucleotides upstream or downstream can influence the rate of transcription. A major difference is that eukaryotes have not been demonstrated to have polycistronic messages.

11. The *araC* product has two conformations, which are determined by the presence and absence of arabinose. When it has bound arabinose, the *araC* product can bind to the initiator site (*araI*) and activate transcription. When it is not bound to arabinose, the *araC* product binds to both the initiator (*araI*) and the operator (*araO*) sites, forming a loop of the intermediary DNA. When both sites are bound to the *araC* product, transcription is inhibited. The *araC* product is trans-acting.

Many eukaryotic trans-acting protein factors also bind to promoters, enhancers, or both that are upstream from the protein-encoding gene. These factors are required for the initiation of transcription. Additionally, some bind to other proteins, such as RNA polymerase II, in order to initiate transcription. Like their counterparts in the *ara* operon, the eukaryotic trans-acting protein factors can bind DNA at two sites, with the intermediary DNA forming a loop between the binding sites.

12. Normally, the repressor searches for the operator by rapidly binding and dissociating from nonoperator sequences. Even for sequences that mimic the true operator, the dissociation time is only a few seconds or less. Therefore, it is easy for the repressor to find new operators as new strands of DNA are synthesized. However, when the affinity of the repressor for DNA and operator is increased, it takes too long for the repressor to dissociate from sequences on the chromosome that mimic the true operator, and as the cell divides and new operators are synthesized, the repressor never quite finds all of them in time, leading to a partial synthesis of ß-galactosidase. This explains why in the absence of IPTG there is some elevated ß-galactosidase synthesis. When IPTG binds to the repressors with increased affinity, it lowers the affinity back to that of the normal repressor (without IPTG bound). Then, the repressor can rapidly dissociate from sequences in the chromosome that mimic the operator and find the true operator. Thus, ß-galactosidase is repressed in the presence of IPTG in strains with repressors that have greatly increased affinity for operator. In summary, because of a kinetic phenomenon, we see a reverse induction curve.

13. Construct a set of reporter genes with the promoter region, the introns, and the region 3´ to the transcription unit of the gene in question containing different alterations that do not disrupt transcription or processing. Use these reporter genes to make transgenic animals by germ-line transformation. Assay for expression of the reporter gene in various tissues and the kidney of both sexes.

15 Regulation of Cell Number: Normal and Cancer Cells

1. **a.** Cyclins bind to, activate, and direct CDKs to phosphorylate specific cellular targets and by doing so control the cell cycle. Normal cell division requires the sequential production and then removal of different cyclins. The activity of the cyclin-CDK heterodimer is also regulated through p21. Overproduction of one of the cyclins could disrupt the orderly process of cell division, but it would be limited by the amount of CDK present as well as the state of the p21 "brake."

   b. A nonsense mutation would lead to a decrease of the normal protein product. If that protein were part of the receptor for a growth factor, which stimulates cell proliferation, then cell division could not be triggered. This would likely be recessive and would slow cell proliferation, not accelerate it.

   c. Overproduction of FasL will signal adjacent cells through their Fas cell-surface receptors, which in turn leads to Apaf activation. This in turn causes proteolysis and activation of the initiator caspase, ultimately leading to apoptosis of the cell. Although this would be dominant, it would lead to excess cell death, not proliferation.

   d. Cytoplasmic tyrosine-specific protein kinase phosphorylates proteins in response to signals received by the cell. These phosphorylations lead to activation of the transcription factors for the next step in the cell cycle. If the active site is disrupted, then phosphorylations will not occur and transcription factors for the next step will not be activated. This would likely be recessive and would slow cell proliferation, not accelerate it.

e. If the enhancer causes large amounts of the apoptosis inhibitor to be expressed in the liver, the normal pathway of cell death will be blocked. These liver cells (the enhancer is liver-specific) will have an unusually long lifetime in which to accumulate various mutations that could lead to cancer. This chromosomal rearrangement would be dominant.

2. Once apoptosis is initiated, a self-destruct switch has been thrown: endonucleases and proteases are released, DNA is fragmented, and organelles are disrupted. This obviously is a "terminal" state from which the cell will not have a need or chance to reuse the machinery of destruction. On the other hand, the various proteins needed for the regulation and execution of the cell cycle will be needed again if the cell continues to divide. By recycling many of these, the cell obviously conserves its resources (proteins are energetically expensive to make) and recycling also allows for more rapid divisions, since the cell does not have to spend time remaking all the pieces.

3. a. This would be dominant. The misexpression of FasL from one allele would be dominant to the normal expression of the wild-type FasL allele. In this case, each liver cell would signal its neighboring cells to undergo apoptosis.

b. No. It would lead to excess cell death, not proliferation.

4. (1) Certain cancers are inherited as highly penetrant simple Mendelian traits.
(2) Most carcinogenic agents are also mutagenic.
(3) Various oncogenes have been isolated from tumor viruses.
(4) A number of genes that lead to the susceptibility to particular types of cancer have been mapped, isolated, and studied.
(5) Dominant oncogenes have been isolated from tumor cells.
(6) Certain cancers are highly correlated to specific chromosomal rearrangements.

5. Normal Ras is a G-protein that activates a protein kinase, which in turn phosphorylates a transcription factor. If it were simply deleted, no cancer could develop, because cell division would not occur. If it were simply duplicated, an excess of the G-protein could not cause cancer, because it must be activated before it can activate the protein kinase, and presumably the enzyme that activates normal Ras is closely regulated and would not activate too many copies. However, if it were to have a point mutation, it might now bind GTP, even in the absence of normal control signals and be in a state of permanent activation. As a positive regulator of cell growth, this mutant Ras would continually promote cell proliferation.

In contrast, normal *c-myc* is a transcription factor. If the gene were to be duplicated, too much transcription factor could lead to malignancy.

6. Inhibition of apoptosis can contribute to tumor formation by allowing cells to have an unusually long lifetime in which to accumulate various mutations that lead to cancer. Also, the normal role of apoptosis in removing abnormal cells and, through p53, killing cells that have "damaged" DNA would also be inhibited.

7. **a.** Mutations in a tumor suppressor gene are recessive and due to loss of function. That function can be restored by the introduction of a wild-type allele.

   b. Mutations in an oncogene are dominant and due to gain of function (overexpression or misexpression). The normal function will not inhibit these mutants, and the introduced gene would be ineffective in restoring the normal phenotype.

8. **a.** Type A diabetes is most likely due to a defect in the pancreas. The pancreas normally makes insulin, and type A diabetes can be treated by supplying insulin. Type B diabetes is most likely due to a target cell defect because type B is unresponsive to exogenous insulin.

   b. Type B diabetes appears to be caused by a defect in the target cell. A number of genes are responsible for the receptor and the subsequent cascade of changes that occur in leading to a change in transcription. Any of these genes could have a mutant form.

9. p53 detects and is activated by DNA damage. When activated, p53 activates p21, an inhibitor of the cyclin-CDK complex necessary for the progression of the cell cycle. If the DNA damage is repairable, this system will eventually deactivate p53 and allow cell division. However, if the damage is irreparable, p53 would stay active and would activate the apoptosis pathway, ultimately leading to cell death. It is for this reason that the "loss" of p53 is often associated with cancer.

16 Pathways of Differentiation

1. Sex determination in *Drosophila* is autonomous at the cellular level. The *Sxl* gene is permanently turned on or remains off early in development in response to the concentration of X:A transcription factor. Because the X:A ratio is established by the interaction of gene products made in the ovary and in the early zygote, a chemical gradient would be expected to exist that would be sufficiently high in some cells to result in femaleness but low enough in other cells to result in maleness, making the individual an intersex.

2. Normally, the *tra* gene in the female is active, while in the male it is not active. The active *tra* form of the gene product results in a change in the *dsx* product, shifting development toward the female. If the *dsx* product is not altered, development proceeds along the male line. A mutation in the gene that results in chromosomal females developing as phenotypic, but sterile, males must involve an inactive *tra* product. Homozygotes for the *tra* mutation could be transplanted with male germ cells very early in development, which should result in normal gonad development.

3. In humans, a single copy of the Y chromosome is sufficient to shift development toward normal male phenotype. The extra copy of the X chromosome is simply inactivated. Both mechanisms seem to be all-or-none rather than to be based on concentration levels.

4. Because maleness is based on the presence of androgens produced by the developing testes and femaleness is based on the absence of those androgens, what seems to be crucial here is whether the migrating germ cells organize testes. Although what determines this is unknown, it may be that a minimal

number of XY cells are required to organize a testis. If, in the mosaic, not enough of these cells exist, then development will be female. If a sufficient number exist, development will be male.

5. The concentration of *Sxl* is crucial for female development and dispensable for male development. The dominant Sxl^M male-lethal mutations may not actually kill all males but simply produce an excessive amount of gene product so that only females (fertile XX and XY) result. The reversions may eliminate all gene product, resulting in XX (sterile) and XY males. The reversions would be recessive because, presumably, a single normal copy of the gene may produce enough gene product to "toggle the switch" in development to female.

6. **a.** There must be a diffusible substance produced by the anchor cell that affects development of the six cells. The 1° has the strongest response to the substance, and the 3° represents a lack of response due to a low concentration or absence of the diffusible substance.

b. Remove the anchor cell and the six equivalent cells. Arrange the six cells in a circle around the anchor cell. All six cells will develop the same phenotype, which will depend on the distance from the anchor cell.

7. **a.** The results suggest that ABa and ABp are not determined at this point in their development. Also, future determination and differentiation of these cells are dependent on their position within the developing organism.

b. Because an absence of EMS cells leads to a lack of determination and differentiation of AB cells, the EMS cells must be at least in part responsible for AB-cell development, either through direct contact or by the production of a diffusible substance.

c. Most descendants of the AB cells do not become muscle cells when P2 is present; all descendants of the AB cells become muscle cells when P2 is absent. Therefore, P2 must prevent some AB descendants from becoming muscle cells.

8. Because the receptor is defective, testosterone cannot signal the cell and initiate the cascade of developmental changes that will switch the embryo from the "default" female development to male development. Therefore, the phenotype will be female.

9. A number of experiments could be devised. A comparison of amino acid sequence between mammalian gene products and insect gene products would indicate which genes are most similar to each other. Using cloned cDNA sequences from mammalian genes for hybridization to insect DNA would also indicate which genes are most similar to each other.

10. **a.** The anterior 20% of the embryo is normally devoted to the head and thorax regions. The bicaudal phenotype results in the loss of these regions and in the loss of A1 through A3. The gap proteins are responsible for the induction of the pair-rule proteins, which ultimately set the number of segments, and the homeotic proteins, which set the identity of the segments. Obviously, the gap proteins are improperly regulated to produce the bicaudal phenotype.

Normal regulation of the gap proteins is accomplished by differential sensitivity to the differing concentrations of the maternally derived morphogens. Because the anterior portion of the embryo has been removed, high concentrations of the morphogens in these regions have also been removed. This results in the abnormal segment number and identity that is observed.

b. The *oskar* mutation results in the loss of the posterior localization of the *nos* mRNA and protein. Therefore, there is no repression of HB-M translation. The lack of repression of the HB-M transcription factor results in an excess of the HB-M protein. The normal shallow gradient, A to P, is therefore lost. Because the gap genes respond differentially to the BCD:HB-M ratio, no induction of the gap genes occurs, which leads to reduced segmentation. This results simply in a broader head and thorax, and no mirror-image phenotype is possible.

11. If you diagram these results, you will see that deletion of a gene that functions posteriorly allows the next-most anterior segments to extend in a posterior direction. Deletion of an anterior gene does not allow extension of the next-most posterior segment in an anterior direction. The gap genes activate *Ubx* in both thoracic and abdominal segments, whereas the *abd-A* and *Abd-B* genes are activated only in the middle and posterior abdominal segments. The functioning of the *abd-A* and *Abd-B* genes in those segments somehow prevents *Ubx* expression. However, if the *abd-A* and *Abd-B* genes are deleted, *Ubx* can be expressed in these regions.

12. Proper *ftz* expression requires *Kr* in the fourth and fifth segments and *kni* in the fifth and sixth segments.

13. It may be that the wild-type allele in the embryo produces a gene product that can inhibit the gene product of the rescuable maternal-effect lethal mutations, while the nonrescuable maternal-effect lethal mutations produce a product that cannot be inhibited.

Alternatively, the nonrescuable maternal-effect lethal mutations may produce a product that is required very early in development, before the developing fly is producing any proteins, while the rescuable maternal-effect lethal mutations may act later in development when embryo protein production can compensate for the maternal mutation.

14. a. The determination of anterior-posterior portions of the embryo is governed by a concentration gradient of *bcd*. The concentration is highest in the anterior region and lowest in the posterior region. The furrow develops at a critical concentration of *bcd*. As bcd^+ gene dosage (and, therefore, BCD concentration) decreases, the furrow shifts anteriorly; as the gene dosage increases, the furrow shifts posteriorly.

b.

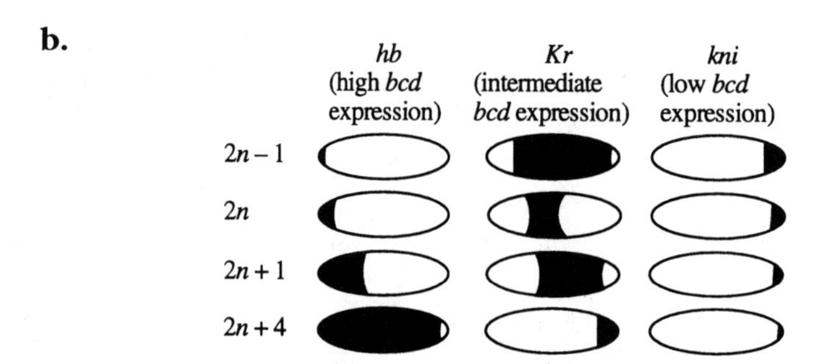

15. The anterior-posterior axis would be reversed.

17 Population and Evolutionary Genetics

1. The frequency of an allele in a population can be altered by natural selection, mutation, migration, nonrandom mating, and genetic drift (sampling errors).

2. There are a total of (2)(384) + (2)(210) + (2)(260) = 1708 alleles in the population. Of those, (2)(384) + 210 = 978 are *A1* and 210 + (2)(260) = 730 are *A2*. The frequency of *A1* is 978/1708 = 0.57, and the frequency of *A2* is 730/1708 = 0.43.

3. The given data are $q^2 = 0.04$ and $p^2 + 2pq = 0.96$. Assuming Hardy-Weinberg equilibrium, if $q^2 = 0.04$, $q = 0.2$ and $p = 0.8$. The frequency of *B/B* is $p^2 = 0.64$, and the frequency of *B/b* is $2pq = 0.32$.

4. This problem assumes that there is no backward mutation. Use the following equation (from *Genetics in Process* 17-3 in the text):

$$p_n = p_o e^{-n\mu}$$

That is, $p_{50,000} = (0.8)e^{-(5 \times 10^4)(4 \times 10^{-6})} = (0.8)e^{-0.2} = 0.65$

5. **a.** If the variants represent different alleles of gene *X*, a cross between any two variants should result in a 1:1 progeny ratio (since the organism is haploid). All the variants should map to the same locus. Amino acid sequencing of the variants should reveal differences of one to just a few amino acids.

   b. There could be another gene (gene *Y*), with five variants, that modifies the gene *X* product post-transcriptionally. If so, the easiest way to distinguish between the two explanations would be to find another mutation in *X* and do a dihybrid cross. For example, if there is independent assortment,

P	X^1 ; Y^1 × X^2 ; Y^2
F_1	1 X^1 ; Y^1:1 X^1 ; Y^2:1 X^2 ; Y^1:1 X^2 ; Y^2

If the new mutation in X led to no enzyme activity, the ratio would be

2 no activity:1 variant one activity:1 variant two activity

The same mutant in a one-gene situation would yield 1 active:1 inactive.

6. **a.** If the population is in equilibrium, $p^2 + 2pq + q^2 = 1$. Calculate the actual frequencies of p and q in the population and compare their genotypic distribution to the predicted values. For this population:

$$p = [406 + {}^1/_2(744)]/1482 = 0.52$$
$$q = [332 + {}^1/_2(744)]/1482 = 0.48$$

The genotypes should be distributed as follows if the population is in equilibrium:

$L^M/L^M = p^2(1482) = 401$	Actual: 406
$L^M/L^N = 2pq(1482) = 740$	Actual: 744
$L^N/L^N = q^2(1482) = 341$	Actual: 332

This compares well with the actual data, so the population is in equilibrium.

b. If mating is random with respect to blood type, then the following frequency of matings should occur:

$L^M/L^M \times L^M/L^M = (p^2)(p^2)(741) = 54$	Actual:	58
$L^M/L^M \times L^M/L^N$ or $L^M/L^N \times L^M/L^M = (2)(p^2)(2pq)(741) = 200$	Actual:	202
$L^M/L^N \times L^M/L^N = (2pq)(2pq)(741) = 185$	Actual:	190
$L^M/L^M \times L^N/L^N$ or $L^N/L^N \times L^M/L^M = (2)(p^2)(q^2)(741) = 92$	Actual:	88
$L^M/L^N \times L^N/L^N$ or $L^N/L^N \times L^M/L^N = 2(2pq)(q^2)(741) = 170$	Actual:	162
$L^N/L^N \times L^N/L^N = (q^2)(q^2)(741) = 39$	Actual:	41

Again, this compares nicely with the actual data, so the mating is random with respect to blood type.

7. **a.**, **b.** For each, p and q must be calculated and then compared with the predicted genotypic frequencies of $p^2 + 2pq + q^2 = 1$.

Population	p	q	Equilibrium?
1	1.0	0.0	Yes
2	0.5	0.5	No
3	0.0	1.0	Yes
4	0.625	0.375	No
5	0.375	0.625	No
6	0.5	0.5	Yes
7	0.5	0.5	No
8	0.2	0.8	Yes
9	0.8	0.2	Yes
10	0.993	0.007	Yes

c. The formulas to use are $q^2 = \mu/_s$ and $s = 1 - W$ (from *Genetics in Process* 17-8 in the text).

$$4.9 \times 10^{-5} = 5 \times 10^{-6}/s\,; \quad s = 0.102, \text{ so } W = 0.898$$

d. For simplicity, assume that the differences in survivorship occur prior to reproduction. Thus, each genotype's fitness can be used to determine the relative percentage each contributes to the next generation.

Genotype	Frequency	Fitness	Contribution	*A*	*a*
A/A	0.25	1.0	0.25	0.25	0.0
A/a	0.50	0.8	0.40	0.20	0.20
a/a	0.25	0.6	0.15	0.0	0.15
				0.45	0.35

$$p' = 0.45/(0.45 + 0.35) = 0.56$$
$$q' = 0.35/(0.45 + 0.35) = 0.44$$

Alternatively, from *Genetics in Process* 17-6 in the text, the formulas to use are

$$p' = p\frac{pW_{AA} + qW_{Aa}}{\overline{W}}$$

$$\overline{W} = p^2W_{AA} + 2pqW_{Aa} + q^2W_{aa}$$

$$p' = (0.5)[(0.5)(1.0) + (0.5)(0.8)]/[(0.25)(1.0) + (0.5)(0.8) + (0.25)(0.6)]$$
$$= (0.5)(0.9)/(0.8) = 0.56$$

8. a. Assuming the population is in Hardy-Weinberg equilibrium and that the allelic frequency is the same in both sexes, we can directly calculate the frequency of the colorblind allele as $q = 0.1$. (Since this trait is sex-linked, q is equal to the frequency of affected males.) Colorblind females must be homozygous for this X-linked recessive trait, so their frequency in the population is equal to $q^2 = 0.01$.

b. There would be 10 colorblind men for every colorblind woman ($^q/_{q^2}$).

c. For this condition to be true, the mothers must be heterozygous for the trait and the fathers must be colorblind ($X^C/X^c \times X^c/Y$). The frequency of heterozygous women in the population will be $2pq$, and the frequency of colorblind men will be q. Therefore, the frequency of such random marriages will be $(2pq)(q) = 0.018$.

d. All children will be phenotypically normal only if the mother is homozygous for the noncolorblind allele ($p^2 = 0.81$). The father's genotype does not matter and therefore can be ignored.

e. There are several ways of approaching this problem. One way to visualize the data, however, is to construct the following:

Mother \ Father	0.4 X^C	0.6 X^c	Y
0.8 X^C	0.32 X^C/X^C	0.48 X^C/X^c	0.8 X^C/Y
0.2 X^c	0.08 X^C/X^c	0.12 X^c/X^c	0.2 X^c/Y

As can be seen, the frequency of colorblind females will be 0.12 and colorblind males 0.2.

f. From analysis of the results in (e), the frequency of the colorblind allele will be 0.2 in males (the same as in the females of the previous generation) and $^1/_2(0.08 + 0.48) + 0.12 = 0.4$ in females.

9. The frequency of a phenotype in a population is a function of the frequency of alleles that lead to that phenotype in the population. To determine dominance and recessiveness, do standard Mendelian crosses.

10. Assume that proper function results from the right gene products in the proper ratio to all other gene products. A mutation will change the gene product, eliminate the gene product, or change the ratio of it to all other gene products. All three outcomes upset a previously balanced system. While a new and "better" balance may be achieved, this is less likely than being deleterious.

11. Wild-type alleles are usually dominant because most mutations result in lowered or eliminated function. To be dominant, the heterozygote has approximately the same phenotype as the dominant homozygote. This will typically be true when the wild-type allele produces a product and the mutant allele does not.

Chromosomal rearrangements are often dominant mutations because they can cause gross changes in gene regulation or even cause fusions of several gene products. Novel activities, overproduction of gene products, etc., are typical of dominant mutations.

12. Prior to migration, $q^A = 0.1$ and $q^B = 0.3$ in the two populations. Since the two populations are equal in number, immediately after migration, $q^{A+B} = {}^1/_2(q^A + q^B) = {}^1/_2(0.1 + 0.3) = 0.2$. At the new equilibrium, the frequency of affected males is $q = 0.2$, and the frequency of affected females is $q^2 = (0.2)^2 = 0.04$. (Colorblindness is an X-linked trait.)

13. For a population in equilibrium, the probability of individuals being homozygous for a recessive allele is q^2. Thus for small values of q, few individuals in a randomly mating population will express the trait. However, if two individuals share a close common ancestor, there is an increased chance of homozygosity by descent, since only one "progenitor" need be heterozygous.

For the following, it is assumed that the allele in question is rare. Thus the chance of both "progenitors" being heterozygous will be ignored.

a. For a parent-sib mating, the pedigree can be represented as follows:

$2pq$

$^1/_2$

$^1/_4$

In this example, it is only the chance of the incestuous parent's being heterozygous that matters. Thus, the chance of the descendent's being homozygous is

$$2pq(^1/_2)(^1/_4) = {}^{pq}/_4$$

If q is very small, then p is nearly 1.0 and the chance of an affected child can be represented as approximately $^q/_4$. (Again, this should be compared to the expected random-mating frequency of q^2.)

b. For a mating of first cousins, the pedigree can be represented as follows:

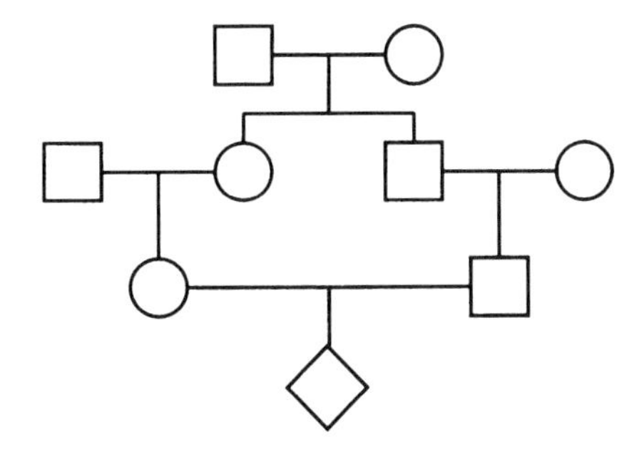

The probability of inheriting the recessive allele if *either* grandparent is heterozygous can be represented as follows:

$2pq$ **or** $2pq$

$^1/_2$ $^1/_2$ $^1/_2$

$^1/_2$ $^1/_2$

$^1/_4$

Thus the chance of this child's being affected is

$$2pq\,(^1/_2)(^1/_2)(^1/_2)(^1/_2)(^1/_4) + 2pq\,(^1/_2)(^1/_2)(^1/_2)(^1/_2)(^1/_4) = {}^{pq}/_{16}$$

Again, if q is rare, p is nearly 1.0, so the chance of homozygosity by descent is approximately ${}^{q}/_{16}$.

c. An aunt-nephew (or uncle-niece) mating can be represented as:

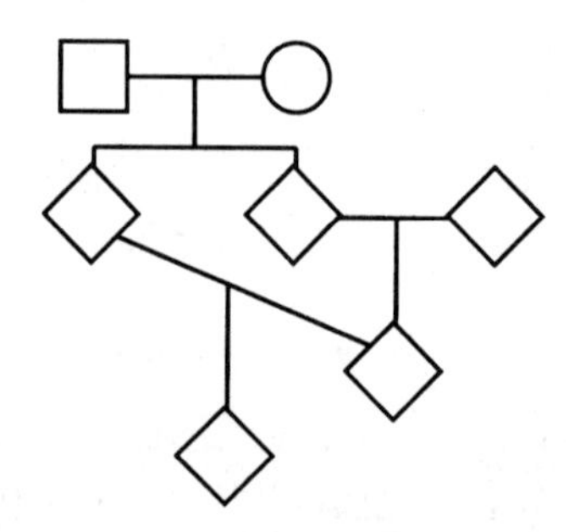

Following the possible inheritance of the recessive allele from either grandparent,

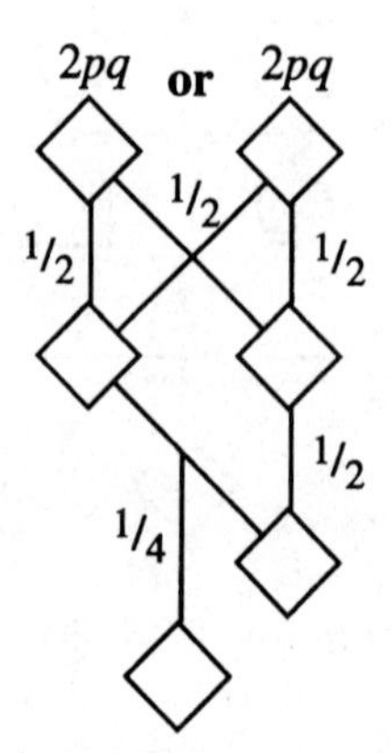

the chance of this child's being homozygous is

$$2pq(^1/_2)(^1/_2)(^1/_2)(^1/_4) + 2pq(^1/_2)(^1/_2)(^1/_2)(^1/_4) = {}^{pq}/_{8},$$

or for rare alleles approximately ${}^{q}/_{8}$.

14. Albinos appear to have had greater opportunity to mate. They may have been considered lucky and encouraged to breed at very high levels in comparison with nonalbinos. They may also have been encouraged to mate with each other. Alternatively, in the tribes with a very low frequency, albinos may have been considered very unlucky and destroyed at birth or prevented from marriage.

15. The allele frequencies are

$$f(A) = 0.2 + {}^1/_2(0.60) = 50\%$$
$$f(a) = {}^1/_2(0.60) + 0.2 = 50\%$$

Positive assortative mating: the alleles will randomly unite within the same phenotype. For $A/-$, the mating population is 0.2 A/A + 0.6 A/a. The allelic frequencies within this subpopulation are

$$f(A) = [0.2 + {}^1/_2(0.6)]/0.8 = 0.625$$
$$f(a) = {}^1/_2(0.6)/0.8 = 0.375$$

The phenotypic frequencies that result are

$A/-$: $p^2 + 2pq = (0.625)^2 + 2(0.625)(0.375) = 0.39 + 0.47 = 0.86$
a/a: $q^2 = (0.375)^2 = 0.14$

However, assuming that all contribute equally to the next generation and this subpopulation represents 0.8 of the total population, these figures must be adjusted to reflect this weighting:

$A/-$: $(0.86)(0.8) = 0.69$
a/a: $(0.14)(0.8) = 0.11$

The *a/a* contribution from the other subpopulation will remain unchanged because there is only one genotype, *a/a*. Their weighted contribution to the total phenotypic frequency is 0.20. Therefore, after one generation, the phenotypic frequencies will be $A/- = 0.69$ and $a/a = 0.20 + 0.11 = 0.31$, and the genotypic frequencies will be $f(A) = 0.5$ and $f(a) = 0.5$. Over time, these allelic frequencies will stay the same, but the frequency of heterozygotes will continue to decrease until there are two separate populations, *A/A* and *a/a*, which will not interbreed.

Negative assortative mating: mating is between unlike phenotypes. The two types of progeny will be *A/a* and *a/a*. *A/A* will not exist. *A/a* will result from all *A/A* × *a/a* matings and half the *A/a* × *a/a* matings. These matings will occur with the following relative frequencies:

$A/A \times a/a = (0.2)(0.2) = 0.04$
$A/a \times a/a = (0.6)(0.2) = 0.12$

Because these are the only matings that will occur, they must be put on a 100% basis by dividing by the total frequency of matings that occur:

A/A × *a/a*: 0.04/0.16 = 0.25, all of which will be *A/a*
A/a × *a/a*: 0.12/0.16 = 0.75, half *A/a* and half *a/a*

The phenotypic frequencies in this generation will be

A/a: 0.25 + 0.75/2 = 0.625
a/a: 0.75/2 = 0.375

In the next generation, since all matings are now between heterozygotes and homozygous recessives, the final allelic frequencies of $f(A) = 0.25$ and $f(a) = 0.75$ will be obtained and the population will be 50% *A/a* and 50% *a/a*.

16. Many genes affect bristle number in *Drosophila*. The artificial selection resulted in lines with mostly high-bristle-number alleles. Some mutations may have occurred during the 20 generations of selective breeding, but most of the response was due to alleles present in the original population. Assortment and recombination generated lines with more high-bristle-number alleles.

Fixation of some alleles causing high bristle number would prevent complete reversal. Some high-bristle-number alleles would have no negative effects on fitness, so there would be no force pushing bristle number back down because of those loci.

The low fertility in the high-bristle-number line could have been due to pleiotropy or linkage. Some alleles that caused high bristle number may also have caused low fertility (pleiotropy). Chromosomes with high-bristle-number alleles may also carry alleles at different loci that caused low fertility (linkage). After artificial selection was relaxed, the low-fertility alleles would have been selected against through natural selection. A few generations of relaxed selection would have allowed low-fertility-linked alleles to recombine away, producing high-bristle-number chromosomes that did not contain low-fertility alleles. When selection was reapplied, the low-fertility alleles had been reduced in frequency or separated from the high-bristle loci, so this time there was much less of a fertility problem.

17. **a.** From *Genetics in Process* 17-6 in the text, the needed equations are

$$p' = p\frac{pW_{AA} + qW_{Aa}}{\overline{W}}$$

$$\overline{W} = p^2W_{AA} + 2pqW_{Aa} + q^2W_{aa}$$

$$p' = 0.5\ [(0.5)(0.9) + 0.5(1.0)]/[(0.25)(0.9) + (0.5)(1.0) + (0.25)(0.7)] = 0.53$$

b. From *Genetics in Process* 17-7 in the text, the needed equation is

$$\hat{p} = \frac{W_{a/a} - W_{A/a}}{(W_{a/a} - W_{A/a}) + (W_{A/A} - W_{A/a})}$$

$$= \frac{0.7 - 1.0}{(0.7 - 1.0) + (0.9 - 1.0)}$$

$$= 0.75$$

18. From *Genetics in Process* 17-8 in the text, the needed equation is

$$q^2 = \mu/s$$

or $$s = \mu/q^2 = 10^{-5}/10^{-3} = 0.01$$

19. Affected individuals = $B/b = 2pq = 4 \times 10^{-6}$. Because q is almost equal to 1.0, $2p = 4 \times 10^{-6}$. Therefore, $p = 2 \times 10^{-6}$.

$$\mu = hsp = (1.0)(0.7)(2 \times 10^{-6}) = 1.4 \times 10^{-6}$$

where h = degree of dominance of the deleterious allele.

20. The probability of not getting a recessive lethal genotype for one gene is $1 - {}^1/_8 = {}^7/_8$. If there are n lethal genes, the probability of not being homozygous for any of them is $({}^7/_8)^n = {}^{13}/_{31}$. Solving for n, an average of 6.5 recessive lethals are predicted.

If the actual percentage of "normal" children is less owing to missed in utero fatalities, the average number of recessive lethals would be higher.

21. **a.** From *Genetics in Process* 17-8, the formula needed is

$$\hat{q} = \sqrt{\mu/s}$$

$$= 4.47 \times 10^{-3}$$

so, Genetic cost $= sq^2 = 0.5(4.47 \times 10^{-3})^2 = 10^{-5}$

b. Using the same formulas as part a,

$$\hat{q} = 6.32 \times 10^{-3}$$

Genetic cost $= sq^2 = 0.5(6.32 \times 10^{-3})^2 = 2 \times 10^{-5}$

c.

$$\hat{q} = 5.77 \times 10^{-3}$$

Genetic cost $= sq^2 = 0.3(5.77 \times 10^{-3})^2 = 10^{-5}$

18 Quantitative Genetics

1. There are many traits that vary more or less continuously over a wide range. For example, height, weight, shape, color, reproductive rate, metabolic activity, etc., vary quantitatively rather than qualitatively. Continuous variation can often be represented by a bell-shaped curve, where the "average" phenotype sis more common than the extremes. Discontinuous variation describes the easily classifiable, discrete phenotypes of simple Mendelian genetics: seed shape, auxotrophic mutants, sickle-cell anemia, etc. These traits show a simple relationship between genotype and phenotype.

2. **a.** Broad heritability measures that portion of the total variance that is due to genetic variance. The equation to use is:

H^2 = the genetic variance/phenotypic variance

where genetic variance = phenotypic variance – environmental variance

$$H^2 = \frac{s_p^2 - s_e^2}{s_p^2}$$

Narrow heritability measures that portion of the total variance that is due to the additive genetic variation. The equation to use is

$$h^2 = \frac{\text{additive genetic variance}}{\text{additive genetic variance + dominance variance + environmental variance}}$$

$$h^2 = \frac{s_a^2}{s_a^2 + s_d^2 + s_e^2}$$

Shank length:

$$H^2 = (310.2 - 248.1)/(310.2) = 0.200$$
$$h^2 = 46.5/(46.5 + 15.6 + 248.1) = 0.150$$

Neck length:

$$H^2 = (730.4 - 292.2)(730.4) = 0.600$$
$$h^2 = 73.0/(73.0 + 365.2 + 292.2) = 0.010$$

Fat content:

$$H^2 = (106.0 - 53.0)/(106.0) = 0.500$$
$$h^2 = 42.4/(42.4 + 10.6 + 53.0) = 0.400$$

b. The larger the value of h^2, the greater the difference between selected parents and the population as a whole and the more that characteristic will respond to selection. Therefore, fat content would respond best to selection.

c. The formula needed is

$$\text{Selection response} = h^2 \times \text{selection differential}$$

Therefore, selection response = (0.400)(10.5% – 6.5%) = 1.6% decrease in fat content, or 8.9% fat content.

3. **a.** The probability of any gene's being homozygous is $^1/_2$ (e.g., for *A*: *A/A* or *a/a*), and the probability of being heterozygous (or not homozygous) is also $^1/_2$. Thus, the probability for any one gene's being homozygous while the other two are heterozygous is $(^1/_2)^3$. Since there are three ways for this to happen (homozygosity at *A* or at *B* or at *C*), the total probability is

$$p(\text{homozygous at 1 locus}) = 3(^1/_2)^3 = {^3/_8}$$

The same logic can be applied to any two genes' being homozygous

$$p(\text{homozygous at 2 loci}) = 3(^1/_2)^3 = {^3/_8}$$

There are two ways for all three genes to be homozygous, so

$$p(\text{homozygous at 3 loci}) = 2(^1/_2)^3 = {^2/_8}$$

b. $p(\text{0 capital letters}) = p(\text{all homozygous recessive}) = (^1/_4)^3 = {^1/_{64}}$

$p(\text{1 capital letter}) = p(\text{1 heterozygote and 2 homozygous recessive})$
$= 3(^1/_2)(^1/_4)(^1/_4) = {^3/_{32}}$

$p(\text{2 capital letters}) = p(\text{1 homozygous dominant and 2 homozygous recessive})$
or $p(\text{2 heterozygotes and 1 homozygous recessive})$
$= 3(^1/_4)^3 + 3(^1/_4)(^1/_2)^2 = {^{15}/_{64}}$

$p(\text{3 capital letters}) = p(\text{all heterozygous})$
or $p(\text{1 homozygous dominant, 1 heterozygous, and 1 homozygous recessive})$
$= (^1/_2)^3 + 6(^1/_4)(^1/_2)(^1/_4) = {^{10}/_{32}}$

p(4 capital letters) = p(2 homozygous dominant and 1 homozygous recessive)
or p(1 homozygous dominant and 2 heterozygous)
$= 3(^1/_4)^3 + 3(^1/_4)(^1/_2)^2 = {}^{15}/_{64}$

p(5 capital letters) = p(2 homozygous dominant and 1 heterozygote)
$= 3(^1/_4)^2(^1/_2) = {}^3/_{32}$

p(6 capital letters) = p(all homozygous dominant) = $(^1/_4)^3 = {}^1/_{64}$

4. For three genes there are a total of 27 genotypes that will occur in predictable proportions. For example, there are three genotypes that have two genes that are heterozygous and one gene that is homozygous recessive (*A/a* ; *B/b* ; *c/c*, *A/a* ; *b/b* ; *C/c*, *a/a* ; *B/b* ; *C/c*). The frequency of this combination is $3(^1/_2)(^1/_2)(^1/_4) = {}^3/_{16}$, and the phenotypic score is 3 + 3 + 1 = 7. For all the genotypes possible, the total distribution of phenotypic scores is as follows:

Score	Proportion
3	$^1/_{64}$
5	$^3/_{32}$
6	$^3/_{64}$
7	$^3/_{16}$
8	$^3/_{16}$
9	$^{11}/_{64}$
10	$^3/_{16}$
11	$^3/_{32}$
12	$^1/_{64}$

And the plot of these data will be

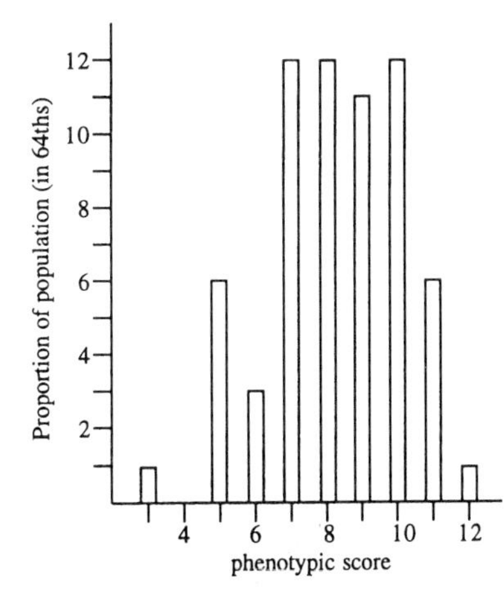

5. The population described would be distributed as follows:

3 bristles	$^{19}/_{64}$
2 bristles	$^{44}/_{64}$
1 bristle	$^1/_{64}$

The 3-bristle class would contain 7 different genotypes, the 2-bristle class would contain 19 different genotypes, and the 1-bristle class would contain only 1 genotype. It would be very difficult to determine the underlying genetic situation by doing controlled crosses and determining progeny frequencies.

6. **a.** Solving the formula for values of x over the stated range for each genotype gives the following data:

x	1	2	3
0.03	0.90		
0.04	0.91		
0.05	0.93		
0.06	0.93		
0.07	0.94		
0.08	0.95		
0.09	0.96		
0.10	0.97		0.90
0.11	0.97		0.92
0.12	0.98	0.90	0.93
0.13	0.98	0.92	0.94
0.14	0.99	0.94	0.95
0.15	0.99	0.95	0.96
0.16	0.99	0.96	0.97
0.17	1.00	0.98	0.98
0.18	1.00	0.98	0.98
0.19	1.00	0.99	0.99
0.20	1.00	1.00	0.99
0.21	1.00	1.00	1.00
0.22	1.00	1.00	1.00
0.23	1.00	1.00	1.00
0.24	0.99	1.00	1.00
0.25	0.99		1.00
0.26	0.99		1.00
0.27	0.98		1.00
0.28	0.98		0.99
0.29	0.97		0.99
0.30	0.97		0.98
0.31	0.96		0.98
0.32	0.95		0.97
0.33	0.94		0.96
0.34	0.93		0.95
0.35	0.93		0.94
0.36	0.91		0.93
0.37	0.90		0.92
0.38			0.90

Plotting these data gives the following curves:

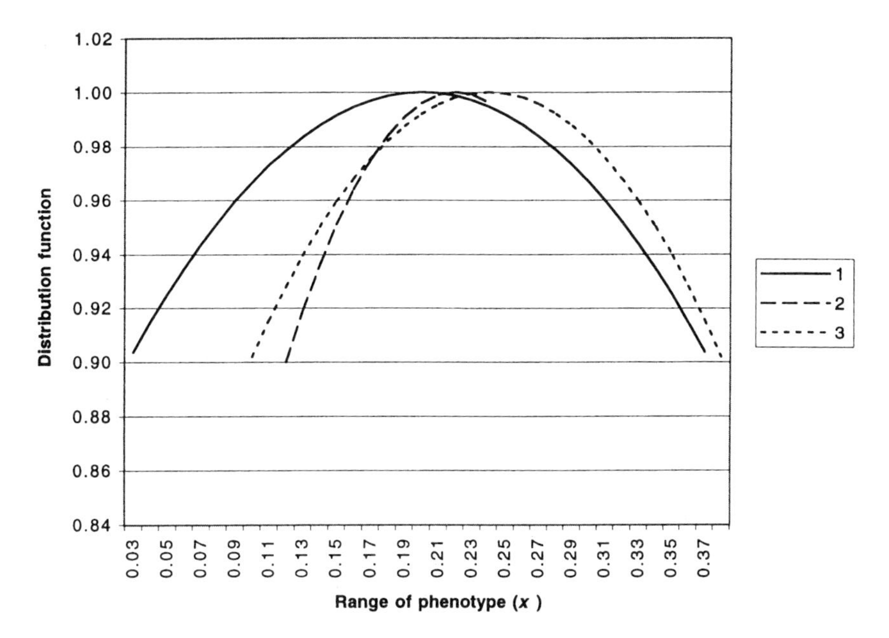

b. Since the three genotypes are equally frequent, the average distribution across the entire range of phenotypes will be

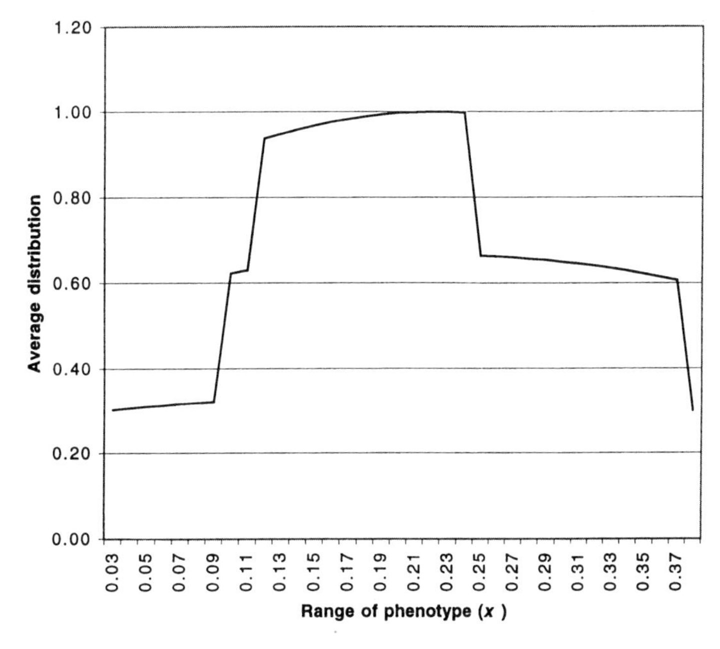

There are regions of the overall phenotypic distribution where the variation within a given genotype does not overlap, and this gives sharp steps to the distribution. On the other hand, any individual whose phenotype lies between values of 0.12 to 0.24 could have any of the three genotypes.

7. The mean (or average) is calculated by dividing the sum of all measurements by the total number of measurements, or in this case, the total number of bristles divided by the number of individuals.

Mean = $\bar{x}$ = [1 + 4(2) + 7(3) + 31(4) + 56(5) + 17(6) + 4(7)]/(1 + 4 + 7 + 31 + 56 + 17 + 4)

= $^{564}/_{120}$ = 4.7 average number of bristles/individual

The variance is useful for studying the distribution of measurements around the mean and is defined in this example as

Variance = s^2 = average of the (actual bristle count – mean)^2

$$s^2 = {}^1/_N \Sigma(x_i - \bar{x})^2$$

$$= {}^1/_{120}[(1 - 4.7)^2 + (2 - 4.7)^2 + (3 - 4.7)^2 + (4 - 4.7)^2 + (5 - 4.7)^2 + (6 - 4.7)^2 + (7 - 4.7)^2]$$

$$= 0.26$$

The standard deviation, another measurement of the distribution, is simply calculated as the square root of the variance:

$$\text{Standard deviation} = s = \sqrt{0.26} = 0.51$$

8. **a.**

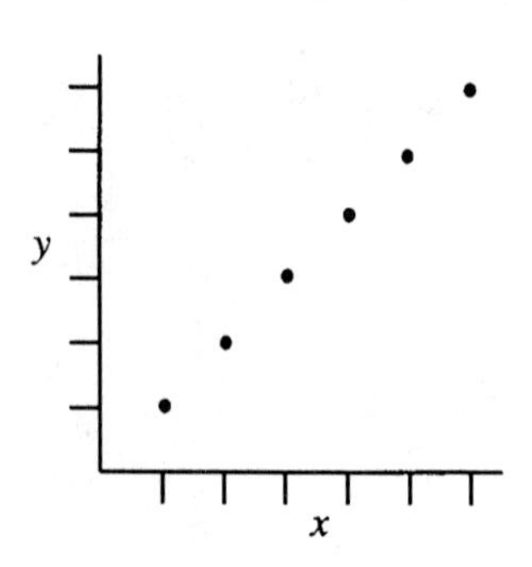

b.

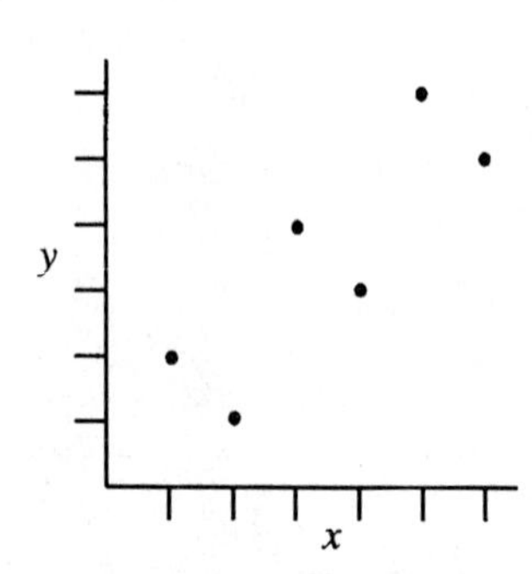

c.

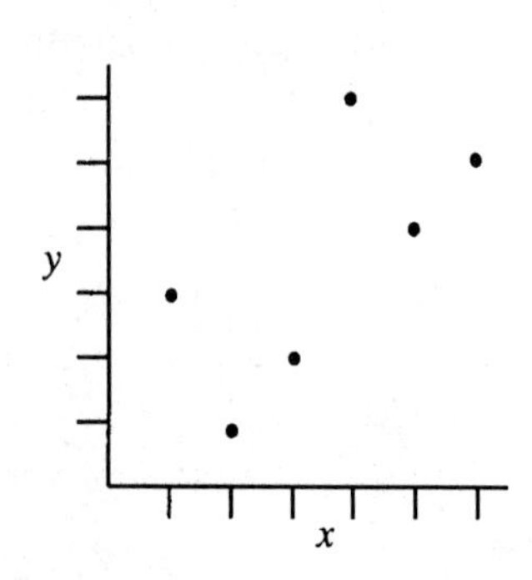

d.

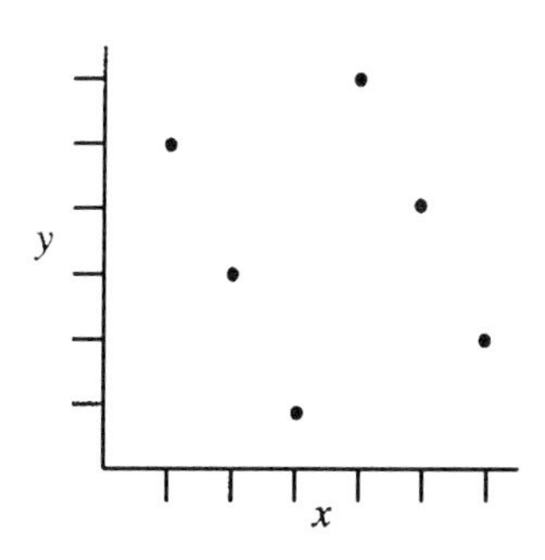

Use the following formulas to calculate the correlation coefficient (r_{xy}) between x and y:

$$r_{xy} = \frac{\text{cov } xy}{s_x s_y} \quad \text{and} \quad \text{cov } xy = {}^1/_N \sum x_i y_i - \overline{xy}$$

a. cov $xy = {}^1/_6$ [(1)(1) + (2)(2) + (3)(3) + (4)(4) + (5)(5) + (6)(6)] – $({}^{21}/_6)({}^{21}/_6)$ = 2.92

$$\text{Standard deviation } x = s_x = \sqrt{{}^1/_N \sum (x_i - \bar{x})^2} = 1.71$$

$$\text{Standard deviation } y = s_y = \sqrt{{}^1/_N \sum (y_i - \bar{y})^2} = 1.71$$

Therefore, r_{xy} = 2.92/(1.71)(1.71) = 1.0. The other correlation coefficients are calculated in a like manner.

b. 0.83

c. 0.66

d. –0.20

9. **a.** H^2 has meaning only with respect to the population that was studied in the environment in which it was studied. Even if a trait shows high heritability, it does not imply the trait is unaffected by its environment. The only acceptable analysis is to study directly the norms of reaction of the various genotypes in the population over the range of projected environments. Since it is so difficult to fully replicate a human genotype so that it might be tested in different environments, there is no known norm of reaction for any human quantitative trait.

b. Neither H^2 nor h^2 is a reliable measure that can be used to generalize from a particular sample to a "universe" of the human population. They certainly should not be used in social decision making (as implied by the terms *eugenics* and *dysgenics*).

c. Again, H^2 and h^2 are not reliable measures, and they should not be used in any decision making with regard to social problems.

10. The following are unknown: (1) norms of reaction for the genotypes affecting IQ, (2) the environmental distribution in which the individuals developed, and (3) the genotypic distributions in the populations. Even if the above were known, because heritability is specific to a specific population and its environment, the difference between two different populations cannot be given a value of heritability.

11. First, define alcoholism in behavioral terms. Next, realize that all observations must be limited to the behavior you used in the definition and all conclusions from your observations are applicable only to that behavior. In order to do your data gathering, you must work with a population in which familiality is distinguished from heritability. In practical terms, this means using individuals who are genetically close but who are found in all environments possible.

12. Before beginning, it is necessary to understand the data. The first entry, h/h h/h, refers to the II and III chromosomes, respectively. Thus, there are four h (high bristle number) sets of alleles in two or more genes on separate chromosomes. The next entry is h/l h/h. Chromosome II is now heterozygous, and chromosome III is still homozygous, etc.

The effect of substituting one l chromosome II for an h chromosome II, and therefore going from homozygous h/h to heterozygous h/l, can be seen in the differences along the rows in the first two columns. The average change is (2.9 + 3.1 + 2.7)/3 = 2.9. When chromosome II goes from heterozygous h/l to homozygous l/l, the average change is (3.2 + 5.2 + 6.8)/3 = 5.1.

The effect of substituting one l chromosome III for an h chromosome III can be seen in the differences between rows: 25.1 – 23.0 = 2.1; 22.2 – 19.9 = 2.3; and 19.0 – 14.7 = 4.3 (average change 2.9). And going from h/l to l/l for chromosome III gives an average change of (11.2 + 10.8 + 12.4)/3 = 11.5 bristles.

Here is a summary of these results:

	Chromosome II	Chromosome III
h/h to h/l	2.9	2.9
h/l to l/l	5.1	11.5

Each set of alleles for both chromosomes is expressed in the phenotype, but that expression varies with the chromosome. Chromosome III appears to have a stronger effect on the phenotype than does chromosome II. (Compare h/h h/h with both l/l h/h and h/h l/l. The difference in the first case is 6.1, and in the second case, 13.3.) Finally, there is partial dominance of h over l for both chromosomes. The change from h/h to h/l is less than the change from h/l to l/l.

13. a.

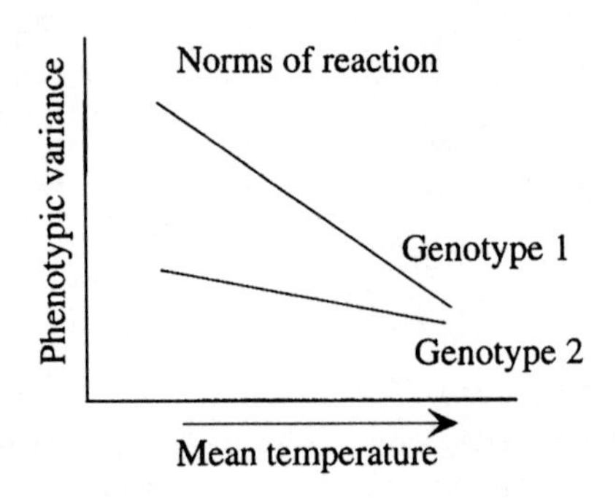

b. Broad heritability is defined as

$$H^2 = \frac{s_g^2}{s_p^2}$$

Assuming that the genetic variance stays the same, phenotypic variance must decrease if H^2 increases. Therefore, the same plot as (a) will satisfy the conditions.

c. To satisfy the conditions that genetic variance increases as H^2 decreases, phenotypic variance must also increase. Therefore the plot will be

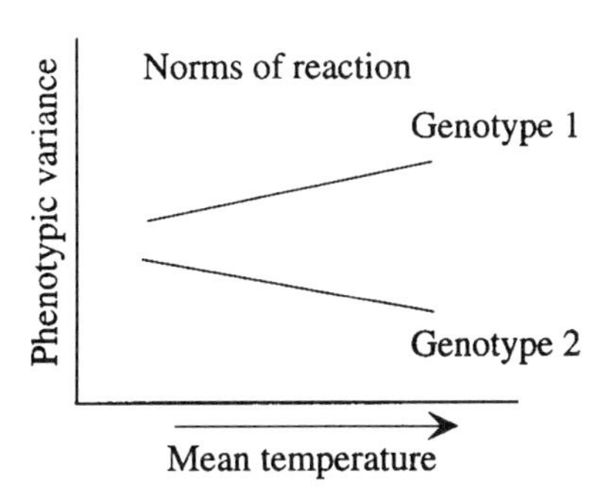

d.

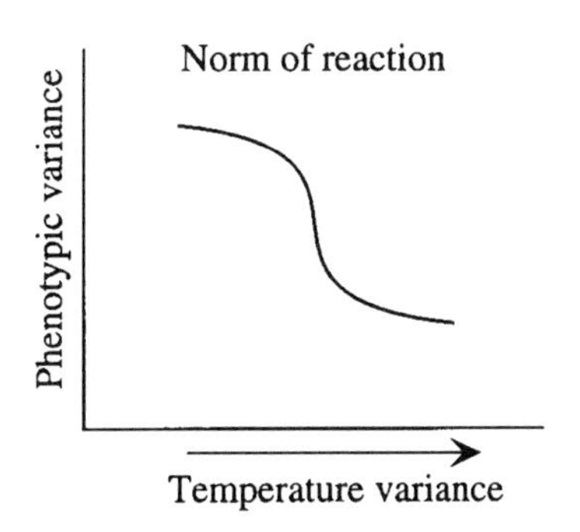

14. a. The regression line shows the relationship between the two variables. It attempts to predict one (the son's height) from the other (the father's height). If the relationship is perfectly correlated, the slope of the regression line should approximate 1. If you assume that individuals at the extreme of any spectrum are homozygous for the genes responsible for these phenotypes, then their offspring are more likely to be heterozygous than are the original individuals. That is, they will be less extreme. Also, there is no attempt to include the maternal contribution to this phenotype.

b. For Galton's data, regression is an estimate of heritability (h^2), *assuming* that there were few environmental differences between all fathers and all sons both individually and as a group. However, there is no evidence given to determine if the traits are familial but not heritable. These data would indicate genetic variation only if the relatives do not share common environments more than nonrelatives do.